East European Change and Shipping Policy

Originally published in 1996, this work begins by considering the changes that have taken place in the social, political and economic environment of Eastern Europe as a whole and then concentrates upon the shipping market with reference to Poland, Romania, Bulgaria, the Czech Republic and Hungary. A detailed model of the relationship between the shipping industry and the contextual changes that have occurred in the region in recent years is then developed before looking specifically at the liner shipping market of Poland and its market positioning within the context of competition in the North Atlantic and European Union operators.

T0143436

East European Change and Shipping Policy

Gillian Ledger and Michael Roe

Routledge
Taylor & Francis Group

First published in 1996
by Avebury (Ashgate Publishing Limited)

This edition first published in 2022 by Routledge
2 Park Square, Milton Park, Abingdon, Oxon, OX14 4RN

and by Routledge
605 Third Avenue, New York, NY 10017

Routledge is an imprint of the Taylor & Francis Group, an informa business

© 1996 G. Ledger and M. Roe

Publisher's Note
The publisher has gone to great lengths to ensure the quality of this reprint but points out that some imperfections in the original copies may be apparent.

Disclaimer
The publisher has made every effort to trace copyright holders and welcomes correspondence from those they have been unable to contact.

A Library of Congress record exists under LCCN: 96084611

ISBN: 978-1-032-14702-4 (hbk)
ISBN: 978-1-003-24065-5 (ebk)
ISBN: 978-1-032-14708-6 (pbk)

Book DOI 10.4324/9781003240655

East European Change and Shipping Policy

GILLIAN LEDGER AND MICHAEL ROE
Institute of Marine Studies
University of Plymouth

Avebury

Aldershot • Brookfield USA • Hong Kong • Singapore • Sydney

Published by
Avebury
Ashgate Publishing Limited
Gower House
Croft Road
Aldershot
Hants GU11 3HR
England

Ashgate Publishing Company
Old Post Road
Brookfield
Vermont 05036
USA

British Library Cataloguing in Publication Data

Ledger, Gillian
    East European change and shipping policy
    1. Shipping - Europe, Eastern
    I. Title II. Roe, Michael, 1954-
    387'.00947

    ISBN 1 85972 384 5

Library of Congress Catalog Card Number: 96-84611

Printed and bound by Athenaeum Press, Ltd.,
Gateshead, Tyne & Wear.

Contents

Figures and tables

Preface

A number of points should be considered before reading this text. Notably, the research forming its basis is extremely diverse, involving a number of areas which have seen continuous change over the period; hence the use of a cut off date to ensure consistency. Significantly, during the course of the research the European Community changed its name to the European Union; to avoid confusion this body is referred to as the European Community throughout the thesis. Also, the Trans-Atlantic Agreement, which was established during the course of the research, has since been both outlawed by the European Commission and replaced by the Trans-Atlantic Conference Agreement (TACA). By the end of 1994 neither the European Commission nor the Federal Maritime Commission of the United States had passed any final decision on TACA, although shipowners were confident that the concessions made would be enough to secure European Commission approval of the Agreement.

1 Introduction

The recent dramatic changes which have taken place in East Europe have created a situation of a scale never before experienced. Social, economic and political changes have affected the activities, organisation and objectives of almost every industry at every level.

This research looks specifically at the impacts of these changes upon the shipping industry within East Europe, and the consequential impacts for its competitors from the European Community. That a direct and substantial relationship between East European social, political and economic changes and shipping interests exists is undeniable, especially where issues such as industrial privatisation, the introduction of bankruptcy laws where none before existed, the moves towards the hardening of currencies and the increased role of market forces are concerned (Lloyds Ship Manager 1991, The Wall Street Journal Europe 1991).

The study of these relationships has particular interest because it is unquestionably unique, topical and dynamic. In an area where there has been little if any previous research, this original work has had to deal with a rapidly and substantially changing industry in a continually developing market.

The research was carried out in direct collaboration with Polish Ocean Lines - the state owned operator which controls the overwhelming majority of Poland's liner shipping - and in close contact with other major organisations involved in East European shipping. Visits were made to all the main maritime operators in Poland, and to several maritime operators and organisations in Belgium and the United Kingdom with interests in East European shipping. Further data and opinion was provided by written contacts throughout the world. Additional information was collected from extensive searches in the UK's major maritime and East European libraries in London, Glasgow, Liverpool, Birmingham, Exeter, Southampton and Plymouth.

There are numerous and substantial problems which one could expect to encounter within this study area. Firstly, the subject area covers a wide range of disciplines dominated by the fields of economics, politics and management but having to incorporate far more. This stems from its attempt to encompass the broad impacts of East European change on the maritime industries of East Europe and subsequently those of the European Community. Each of these areas must be researched, understood and integrated within the thesis.

Secondly there is a serious lack of reliable data concerning all issues of East Europe; much of the information produced under the old Communist regime is very unreliable, and the amounts produced both today and in the past are relatively limited and fragmented.

Further problems arise due to the amount of time taken for data, when reliable, to emerge from East Europe, and for correspondence to pass through a poor communications infrastructure, and from a management culture in the East which is unused to collaboration with the West. Additionally, a constantly changing political and economic environment means that information quickly becomes outdated - exemplified by the creation of new states during the course of the study, such as Slovakia and the Czech Republic.

There are drawbacks due to the limitations of previous work in the area of the East European maritime industries, although there are some relatively outdated reports, for example by ESCEC (1977) and Bergstrand and Doganis (1987). However, the lack of previous work in the area adds to the qualities of originality, uniqueness and interest. Meanwhile, working in the area of East Europe and in an industry which operates on an international basis may also create problems concerned with language barriers.

Finally, the competitive and commercial nature of the maritime industry throughout Europe means that shipping companies are not always willing to divulge large quantities of useful data and opinion.

Given this range of problems, the broad objective of this research is to examine the relationship between East European change and shipping, and more specifically has been restricted to the following:

1   To assess the situation of the Polish shipping industry before and after the changes which have taken place, between 1988 and 1992.

2   To analyse the effect of East European change upon Polish shipping and to develop models to identify the major areas of change and the relationship of changes to the shipping industry.

3   To analyse through further model development, the significance of change in Polish shipping when related to European Community based companies operating in a competitive market.

The research area has been constrained geographically to exclude the ex-Soviet Union. Reasons for this centre upon the uncertain future condition of the ex-Soviet Union and the fact that the shipping industry there is already relatively well researched and documented (Bergstrand and Doganis 1987, Ambler, Shaw and Symons 1985, Guzhenko 1977, Lloyds List 11/10/91, Lloyds Ship Manager 1991 and 1992, Maslov 1984). Also, Hungary has been excluded due to its relative insignificance, a lack of available information and lack of cooperation by Hungarian shipping interests. With constant changes taking place, the research also had to be constrained to a certain time period. The end of 1992 was used as a cut off date to provide consistency throughout the study. Consequently, detailed social, economic and political changes since then have been largely excluded.

Before the results of these changes can be examined in any detail it is first necessary to complete a review of the recent history of East Europe, and it is to this that we now turn our attention.

2 East European change and shipping

This study of the relationship between East European shipping and social, political and economic change in that region has been constricted to the countries of Poland, Bulgaria, the Czech Republic, Slovakia, and Romania. The ex-Soviet Union has been excluded from the main study due to its considerable complexity and continued chaotic condition, Hungary because of its limited maritime interests, and East Germany has been excluded as reunification in 1989 meant that it automatically became part of the European Community and was thus placed in a situation altogether different from the other East European countries.

Any assessment of developments in the shipping industry in East Europe has to begin with an interpretation of the broad economic and political structures that dominated before the changes of the late 1980's. The Council for Mutual Economic Assistance (CMEA), also known as COMECON, was East Europe's closest equivalent to the European Community. It consisted in its final days of the Soviet Union, Poland, Czechoslovakia, Hungary, East Germany, Romania, Bulgaria, Cuba, Mongolia, Vietnam, and at times in the past Associate Members, such as Albania and Yugoslavia (Financial Times 10/05/91). It was associated in January 1949 as a response to the Marshall Plan which provided United States finance to repair physical and economic war damage in Western Europe.

The official, although at times probably theoretical, tasks of the CMEA were:

1   to provide a forum to exchange information on trade,
2   to co-ordinate trade between member states and with the outside world,
3   to provide the mechanism for exchange of goods within the CMEA without involving monetary movements of hard currency, largely based around the use of the transferable rouble.

Unofficially, but in practice, the CMEA provided a way for the Soviet Union to control the satellite states of Eastern Europe and to dictate the latter's production to feed the Soviet economy (Europa 1994). This system was imposed by the Soviet Union militarily and also economically by insisting that payment for goods from satellites was either in barter form or in transferable roubles which were largely unusable except in exchange for more Soviet goods, thus tying East European economies over the long term to that of the Soviet Union (The Independent 29/06/91).

The CMEA had no constitution and supposedly provided an informal mechanism; in theory any member state could ignore CMEA requirements and decisions and could leave at any time. In practice members were economically and militarily bound into the Soviet led system and rarely ignored its requirements, with for example, open refusal to conform to its decisions occurring only three times in its forty one years existence.

The CMEA worked on two main economic and politically inspired principles aimed at strengthening the linkages between member states and the Soviet Union in particular (Drabnek 1989).

1. The planned economy

In 1962 a planned economy system was introduced for all Member States. This was supposedly economically advantageous because production levels could be co-ordinated with each other to achieve economies of scale and there would also be advantages from linked and co-ordinated plan targets.

Under the CMEA umbrella there were four levels of economic plan (Mieczkowski 1978):

i   Long range broad CMEA plans - up to 15 years.
ii   Plan for the state - for example Poland, normally 10 or 5 years.
iii   Plan for the sector - for example iron and steel, normally covering 5 years.
iv   Industrial plant or enterprise plan - for example an iron and steel works; plans from 3 months to 1 year.

Transport in general, and shipping in particular, was subservient to each of these plan levels. The central focus remained on production targets to which transport was later planned to match, and forced to accept as a priority over passenger movements.

Despite being sound in theory the planned economy was subject to a number of practical criticisms (Financial Times 10/05/91):

i   It was bureaucratically inefficient - it was common for 15% of the Member States' employees to be employed solely in the administration of the planning mechanism.

ii   It was inherently corrupt. The state paid employees a standard wage with a bonus if the enterprise exceeded targets. Many enterprises claimed to have exceeded targets which commonly were not achieved, by providing spurious statistics which went largely unchecked by an administration unable to cope and with little incentive to do so.

iii   There was a failure of co-ordination between the four system levels.

iv   Transport amongst other service activities (eg energy) was treated as subordinate to industrial needs and played no role in creating demand, acting subserviently to industrial production needs.

v   The objectives of the plan were almost universally for quantity not quality. Market demand was guaranteed through the transferable rouble and barter system, and numerous restrictions on international trade and the import of competing products from the west.

2. The Division of International Socialist Labour

The CMEA dictated where different items were to be produced, by assessing the strengths, suitability, and efficiencies of different locations, plants and economies. The principle was logical and suggested potential gains from economies of scale and specialisation. However, a lack of logic in where items were produced and a lack of co-ordination caused major problems. In practice there were no competitors (for example the majority of buses were produced in Hungary); a lack of local production relating to local needs; failure to market differentiate (for example between Romania and the Eastern Soviet Union); which led to a lack of market response, no incentives to compete, and huge, unresponsive, inflexible plants.

Another feature of the former Eastern bloc was the Foreign Trade Organisation (FTO). Large numbers of these existed in each country, responsible for all imports and exports in a particular industrial service or commercial field. No trade could occur without permission from the relevant FTO, which were state controlled monopolies dictating contract details including transport mode and operator for imports and exports; source of imports and exports; and trading terms, for example CIF (Carriage, Insurance and Freight) exports and FOB (Free On Board) imports.

Trade was controlled in such a way, to ensure that the CMEA maximised its potential for earning convertible currency, and to conserve the hard currency that member countries did possess. The currencies of member states were unconvertible and could not be used to purchase much needed foreign goods. The role of shipping, along with other activities such as international haulage, was significant as it operated in an international market and was a primary source of hard currency earnings. The control of shipping operations and in particular trading terms through the FTOs, meant that the majority of sea trade with non-members was carried out by Eastern bloc vessels which either earned hard currency in the process, or conserved the expenditure of existing reserves. Western vessels were used only when no suitable Eastern bloc vessel was available.

Because of the dearth of reliable information from much of the remainder of East Europe the ex-Soviet Union provides the best example of shipping organisation pre-perestroika, that is, until 1985, even though this research will focus on other countries. The same basic organisational principles were applicable to all East European countries before 1985, and to a certain extent remain with us to this day, as we shall see later.

USSR shipping was state run under a central planning system with four key organisations:

i   MORFLOT - the Ministry related to Merchant Marine was set up in 1946. Morflot had to collaborate with the umbrella planning authority GOSPLAN which was responsible for economic policy-making and broad monetary matters throughout all sectors of the USSR.

ii   Shipping Holding Corporations (Sevzaflot, Yuhzflot and Dalflot) -each holding corporation had a budget from Morflot and was related to a certain geographical area. It was a planning organisation designed to achieve more precise control of the shipping companies.

iii   Shipping Companies - defined geographically; each shipping company received a budget from one of the holding corporations. There was a great deal of interaction between the companies and the corporations;- for example, frequent loaning of ships. The shipping companies were also responsible for running the ports.

iv   Other numerous groups and agencies, for example Sovinflot the state owned shipping agency, and Sovfracht the state run chartering monopoly.

It was also widely recognised (Bergstrand and Doganis 1987) that within this organisational structure shipping companies in each CMEA country were involved in three different activities:

i   Genuine commercial activity resembling that of West European market economy shipping companies, to import and export goods and provide services for the cross trades - ie trades that involve countries whose nationalities differ to that of the vessel's flag.

ii   A specific directive of earning convertible currency, which affected, markedly, overall commercial policy. For example cross trading provided a method of earning hard currency whilst domestic trading enabled currency conservation. At times the requirements of convertible currency earning would supersede other commercial needs leading, for example, to pricing at lower than cost, simply to earn DM/US$ etc.

iii   Providing logistical support. Most Eastern bloc shipping had an alternative and clandestine defence function.

In summary, the CMEA was Soviet dominated and created broad plans for member states, consequently and ultimately controlling the production levels of each enterprise. Shipping was then subservient to production and was also subject to the actions of other state controlled groups such as shipping and chartering agencies and those involved in maritime organisation, finance, and planning. The system was centrally operated with enterprises reacting to central needs rather than to markets. Decision making therefore was downwards (from central government), rather than upwards (from the consumer) as in a more western style economy.

With the accession of Gorbachev to the Presidency of the Soviet Union, decreasing Soviet intervention in other East European states, declining East European economies, lack of hard currency, and pressure from the European Community, United States, International Monetary Fund and the World Bank, the CMEA eventually collapsed (Financial Times 24/07/91). On 28th June 1991, at a meeting of its nine member countries in Budapest, the CMEA was formally dissolved with effect in 90 days (Financial Times 29/06/91).

Effectively the CMEA had ceased trading two years previously. Its members realised that it was 'economic nonsense' to have an international pricing system completely removed from world markets (Independent 29/06/91). They began to look to the West for trading partners that would pay in hard currency, vital to East European economies. According to Bela Kadar, Hungarian Minister of International Economic Relations and host of the organisations represented at the 46th and closing session, 'Comecon was unable to give answers to the challenges of a changing world'. The Soviet news agency Tass reported that 'The CMEA has exhausted itself but the economic space of its member countries is not a myth but a reality. This must be taken into account, and member countries should not break the traditional economic links between them'. Pravda described the organisation's 'bulky' bureaucracy as 'legendary' (Financial Times 10/05/91).

Comecon's only successor was a liquidation committee, given 90 days to work out the distribution between members of the Moscow headquarters and other common assets. By 24th August 1991 representatives of the nine member states had not managed to agree on a value for common assets, although the issue had to be resolved by 28th September, the date for the final liquidation of the organisation (Financial Times 24/07/91).

The effects of the demise of the CMEA were reported in the Economist on the 20th April 1991. As organised barter and transferable roubles were replaced with trade in US dollars which the indebted East European companies could ill spare, much of it simply stopped. Factories which made the same product year after year for the same undemanding customer were faced with bankruptcy notices - for example dozens of Poland's largest industrial companies faced closure, according to its then Prime Minister Jan Krzysztof Bielecki (The Economist 20/04/91).

It was proposed that the CMEA would be replaced with the 'Organisation for International Economic Co-operation' (OIEC). However, East European governments differed in opinion as to whether any replacement was needed. The Poles and Hungarians preferred the idea of a free trade zone in their part of Europe and would have liked Czechoslovakia to join them in this. Conversely the old USSR, Bulgaria, Romania, Cuba, Vietnam and Mongolia appeared to be in favour of an OIEC (The Economist 20/04/91, Financial Times 24/07/91, The Guardian 20/12/90). The situation remained uncertain for some while until the whole idea of a new CMEA lost favour and was replaced by a marked move by many of the ex-members (notably Poland, Hungary and Czechoslovakia) towards the west and the European Community markets in particular. During 1992 a free trade agreement was signed between these three countries (now four) but was restricted in scope and authority .

Although this provides a broad picture of some of the significant changes which have been taking place in Eastern Europe, particularly around the period of the demise of the CMEA, it is necessary to look at each country that this research concentrates upon to achieve greater detail, and to begin to understand the significance of the social, political and economic changes that have occurred. As the situation is one of continual change, only those developments up to the beginning of 1993 have been included.

Poland

Recent Polish political problems and those of the 1970s and 1980s are the direct consequence of its economic problems during this period, and of the enormous changes in the social, economic and political framework of the country. The over ambitious rate of growth planned by the policy makers of Edward Gierek's Communist government in the early 1970s was led by imports, primarily from the West, which were expected to provide Poland with modern technologies, but the economy failed to generate the production levels and the exports which had been expected. Imports grew to such volumes that drastic cuts were imposed later in the 1970s. Investment projects were halted. Bad harvests aggravated the problems by making increased food imports necessary. The economic downturn, for many key commodities, started in 1979, well before the unrest associated with Solidarity and the free trade union movement, which led ultimately to the collapse of Communist control and Soviet Union domination (Staar 1993).

Throughout 1980 and 1981 there were riots and strikes as food price rises were introduced. Then the movement for free trade unions arose, out of the workers confrontation with the government and their bid to influence the economic development of the country. Solidarnosz, the Solidarity trade union, was the mass movement which emerged, and it soon had a membership of ten million people. However, a change of government, the imposition of martial law, and the outlawing of Solidarity brought about a calmer and more ordered situation - if one rather more politically and democratically unsatisfactory.

Throughout 1981 economic output slumped. In fact, the 1981 production figures are excluded from the official government statistics series, omitting the record of a disastrous year. The poor economic performance was aggravated by the 'free Saturdays' which Solidarity had won for workers, giving them a five day week.

In September 1981 Solidarity held its first (and last) party congress. It stated that the only way to achieve economic recovery was through the acceptance and implementation of the principle of worker self-management. It urged workers in other East European countries to set up their own independent unions. It demanded basic economic reform, free elections and the right to broadcast - the latter had been promised in the Gdansk agreement with state politicians in September 1980, but had not been implemented.

In October 1981, General Jaruzelski replaced Kania as First Secretary of the party. In December 1981 he set up the Military Council for National Salvation, which was to run the country under twelve generals and which was to put a stop to processes which were deepening Poland's economic crisis. Martial law was declared and protestors were severely dealt with, some being shot dead. Strikes were declared illegal. Certain sectors of the economy were placed under military discipline. Solidarity leader Lech Walesa was interned, whilst Solidarity itself was banned in October 1982. Walesa was later awarded the 1983 Nobel Peace Prize, despite intensive official attempts to discredit him. Opposition to the Government continued through a network operating underground, producing books, bulletins and broadcasts (Koralewicz, Bialecki and Watson 1987) .

The government's programme was gradually implemented and output levels improved from the second half of 1982. New, government sponsored trade unions were set up, which by late 1983 had attracted 3.5 million members. A new political association, the Patriotic Movement for National Rebirth (PRON) was formed, and given a constitutional role. The church's influence was used to promote new organisations.

Martial law was suspended in December 1982 and then lifted in July 1983, but disturbances continued with many clashes between police and demonstrators. These incidents were played down by the media, whilst repression continued to affect many people. Despite being banned, Solidarity still operated, its supporters collecting subscriptions to sustain the families of members in prison, or those unable to find work for political reasons.

The government then began to pursue a more conciliatory approach. Charges against Solidarity activists awaiting trial were reduced, while some dissidents were offered freedom if they agreed to emigrate (East Europe Economic Handbook

10

1985).

The policy adopted by the Jarulzelski government was not unproductive. In an amnesty during September 1986 virtually all political prisoners were released and moves were made within days to set up a provisional council of Solidarity, while nevertheless preserving its underground leadership (the TKK). By taking this initiative the Jarulzelski leadership caused a 'great deal of confusion and misunderstanding among members of Solidarity' (Joint statement by Walesa and TKK 1986).

Measures such as the 1986 amnesty did not reflect a major change in the regime's approach. Nevertheless, the authorities took care to show considerable sensitivity in their handling of relations with the church (Rzeczpospolita 01/06/87). Additional evidence of the leadership's desire to establish further relations with independent social forces came with the unexpected convocation of a Central Committee meeting in October 1987 and a decision to hold a referendum over further reform proposals. However, the result of the referendum, with government proposals failing to get the approval of more than 50% of those eligible to vote, did little to improve the relations of the regime with Polish society (Kolankiewicz and Lewis 1988).

In the traditional Communist style, the Polish economy had been managed in a centrally controlled manner, although the degree of control had been progressively relaxed in the 1980s in parallel with political developments, allowing for example, increased private economic activity.

In 1989 Poland became the first former satellite to extricate itself from Soviet domination and vote out the Communists in relatively free elections. Since then continual disagreements have caused four governments to collapse in as many years (Blazyca and Rapacki 1991).

At the end of 1989 the new Polish government introduced major new policies directed towards liberalising the economy, introducing market principles, privatising the state's assets, stabilizing the economy (particularly with respect to inflation) and making the zloty convertible for Polish individuals and organisations, although not fully convertible on the international markets.

In line with other East European countries Poland progressively adopted a market led political and economic policy for all sectors which resulted ultimately in considerable additional freedoms for organisations and individuals. The economy suffered as a result, for example through inflation, production slumping by forty percent, and unemployment climbing from zero to thirteen percent in three years (Financial Times 24/02/93); however, the process of democracy and Communist state withdrawal had begun in earnest. By the 10th September 1991, Lloyds List reported that Poland had moved away from martial law and hard line Communist government and had 'embraced capitalism quickly and enthusiastically' (Lloyds List 10/09/91).

On 28th October 1991, Poland held its first significantly free parliamentary election since the thirties. Poland's Democratic Union Party, led by former Prime

Minister Tadeusz Mazowiecki emerged with the biggest share of the vote. However, 60% of the electorate signalled that it had had enough of arguing politicians and falling living standards, by boycotting the elections. As a result Poland was divided politically between a number of small parties with little prospect of forming a stable parliamentary coalition able to sustain a government with an agreed programme (Financial Times 29/10/91).

During September 1992 the World Bank released a study of the Polish economy which stated that 'the fiscal situation has worsened to the point where hyperinflation is an immediate danger. Unemployment has reached a level that cannot be tolerated' (World Bank 1992). At the same time Fairplay (03/09/92) highlighted the enormous scale of debt that Poland had inherited. State enterprises both large and small were notorious debtors, being indebted to the state budget, to banks, and to each other.

On 28th May 1993, President Lech Walesa refused to accept Prime Minister Hanna Suchocka's resignation after the no-confidence vote in her government had been approved in the Sejm by a narrow margin. Suchocka's Democratic Union party had grown out of the Solidarity movement and was a keen advocate of economic reforms (The Independent 31/05/93). On the afternoon of May 31st the President announced that he was dissolving Parliament, and called for fresh parliamentary elections although these were not due until 1995 (Business Central Europe 1993, The Economist 18/09/93).

A new election law produced a theoretically more workable parliament in the election of September 19th, in that the previous 29 factions were reduced to six parties. The result of the election was to place parliament in the hands of parties descended from the old communist regime. However, the Economist reported on 25th September 1993 that this did not mean that Poland was reverting to Communism (Economist 25/09/93). The Democratic Left Alliance, which emerged from the ruins of the Communist party, won 173 of parliament`s 460 seats. The Alliance was led by young politicians who claimed to be Social Democrats and preached market reforms and privatisation. They also favoured moderately increased social spending. The other party to fair well with around 128 seats was the Polish Peasant Party. The Labour Union gained around 42 seats, the Democratic Union 69, the Confederation for an Independent Poland 24, and the Non Party Block to Support Reform 20 seats.

In October 1993 International Management noted that while the rest of East Europe was in decline, Poland was recovering after more than three years of former finance minister Leszek Balcerowicz's 1990 'shock therapy programme'. GDP rose by 1.2% in 1992, despite a drought that hit agriculture. Industrial production increased by 4%. During the first half of 1993 output grew by 9% over the same period of 1992. Inflation which was almost 700% in 1990, slowed to 44% in 1992, was below 36% in 1993 and was expected to be below 25% in 1994. Productivity in manufacturing and construction rose by 13% during the first half of 1993 compared with the first half of 1992 (International Management

1993).

For all the problems in the fourth year of Poland's transformation, Dornberg, a Warsaw journalist, writes that the 'good news outweighs the bad' and that there is a 'feeling that the Polish miracle will endure and that the economy is on the launchpad' (International Management 1993).

Romania

In 1948 a 'Treaty of Friendship and co-operation' with Moscow cemented Romania's status as a satellite. In 1951 the 'Moscow Communists' led by Ana Pauker controlled the top government posts. Meanwhile the nationalist Gheorghiu-Dej established power throughout the Romanian Workers Party and became general secretary in 1945, finally toppling Pauker's faction from power in 1952. Stalin's death in 1953 enhanced Gheorghiu-Dej's belief that he alone could determine Romania's path. He firmly believed that his regime's Socialist future depended on the forced growth of heavy industry. Khrushchev's plans for the CMEA involved Romania as a provider of oil and agricultural products, but Gheorghiu-Dej opposed these plans and in the early sixties unveiled plans that would define Romania's independent path over the next three decades. Following Gheorghiu-Dej's death in 1965 Ceausescu took care to preserve Romania's autonomous stance (Zonis and Semler 1992).

It took Ceausescu approximately four years to consolidate his personal control over the renamed Romanian Communist Party. During these years he mirrored his predecessors commitment to industrialisation and an independent foreign policy. He displayed populist tendencies and opened up political debate and participation for the masses. Travelling the country widely he raised the hopes of many Romanians that a more relaxed political era had begun. In foreign policy in 1968 when Soviet tanks rolled into Czechoslovakia, Ceausescu denunciated the invasion and Romania refused to participate. Romania's relations with the USSR deteriorated and it reduced its contacts with the Warsaw Pact.

In 1973 Ceausescu launched the economic campaign that was to define the rest of his reign. During a party meeting he officially redefined Romania as a 'developing socialist country', which gave him the authority to launch an ambitious industrialisation campaign, and enabled him to obtain concessional trade provisions from many western nations. During the 1970s Ceausescu's strategy paid important dividends. Romania's economy grew vigorously and Ceausescu was feted as a courageous statesman. However by the end of the 1970s the plans had gone awry. Bucharest had imported large amounts of industrial goods on credit from the West. Foreign exchange was necessary to pay foreign bills, yet Ceausescu had limited his options by concentrating on expanding the country's petrochemical industries. With Romania's once large oil supplies dwindling, the economy became precariously dependent upon the limited amount of oil that

13

remained. Unless it could import Middle Eastern oil at favourable prices, its petrochemical exports would shrink and the country's financial burden would become unsupportable. The combination of OPEC price shocks, the Iranian revolution, and the Iraq-Iran war proved devastating for the country. Romania's industrial productivity plummeted, and its foreign debt increased rapidly to $16 billion (Nelson 1992).

In 1977 a large scale miners strike in the Jiu Valley signalled growing unrest. Romania's economic crisis contributed to Ceausescu's declining legitimacy. As his economic plans faltered Ceausescu increased political repression to sustain his political power. He believed that in Poland the huge foreign debt had played a significant role in its economic collapse which in turn had forced a change in the country's leadership. Fearing for his own power Ceausescu became obsessed with erasing the debt.

In 1982 he declared that Romania's foreign debt would be eliminated by the end of the decade. He also demanded that the pace of economic growth be maintained. These could be accomplished only if the population's standard of living was severely depressed. Ceausescu succeeded in rapidly reducing the foreign debt, but the rationing of food, basic amenities and energy produced virtual wartime conditions. Exiled dissidents estimated that at least 15,000 Romanians died annually from malnutrition, cold and lack of medical care. Meanwhile Ceausescu and his immediate supporters continued to live in luxury. By 1989, the year of the revolution, Ceausescu was hated by his people, ostracised by fellow socialist leaders and vilified in the west. Still Ceausescu remained certain of his power and crushed dissent wherever it arose, until the Romanian people finally rose up in revolution in 1989 and Ceausescu was killed (Ratesh 1993).

Throughout his rule Ceausescu resisted any retreat from his Stalinist economic model. Even in 1991, two years after the revolution, Romania remained the most centralised command economy in East Europe with an industrial infrastructure in complete decay. By late 1991 per capita gross domestic product was only half that of 1989 East Germany, and consumer goods were generally nonexistent. The new government of the National Salvation Front (NSF) expressed its intention of joining other former satellites in a reorientation to a market economy. However, it is widely agreed that Romania faces the greatest difficulties of all the CMEA countries in transition. Because of Ceausescu's human rights violations and suspicions that his overthrow was less the product of a popular uprising than a long planned coup with at least initial backing from the KGB, the government is largely isolated, both from the newly reforming East and the West. Only by mid 1991 did that isolation begin to break as the IMF entertained the possibility of a $1 billion loan (Zonis and Semler 1992). Since then only very gradual progress has been made towards the development of a less centralised and more liberalised economy.

Bulgaria

Bulgaria was taken over by the Communists on 9th September 1944. One day after the Soviet army had entered Bulgarian territory, the Fatherland Front, the antifascist coalition founded by the Communist party, came to power. The Fatherland front was a coalition of parties of which the Communist Party was the strongest. Until 1945, the Fatherland Front operated as a coalition called the Bulgarian Agrarian National Union (BANU), but Giorgi M. Dimitrov presented his party as an independent alternative to the Communists. Dimitrov resigned under pressure in January 1945 and his removal marked the end of the coalition. The Communist party began choosing the leaders of other parties in the coalition and suppressing its partners. Eventually a series of show trials saw defendants accused as traitors and spies and sentenced to death or life imprisonment.

The Communist Party came to power in Bulgaria through a series of carefully planned moves and received its greatest support from the occupying Red Army. Both the army and the party used brutal force to eliminate its competitors (Zonis and Semler 1992).

The Bulgarians viewed the previous 500 years of Turkish rule as a period of economic stagnation, and after World War II, the immediate goal was to industrialise the country. A two year plan began the process through the implementation of a Stalinist economic model of state owned industry and collectivised agriculture. The plan advanced Bulgaria rapidly as an urbanised and industrialised country, as is evidenced by the following figures:

	Urban Population (% of total)	% of population employed in industry
1930	20.0	7.7
1960	38.0	21.9
1985	64.8	37.2

Source: Zonis and Semler 1992

Heavy industry and industrial employment grew rapidly, but neither agricultural employment nor living standards matched the gains of heavy industry. When Stalin died in 1953 there was a move for reform.

In 1954 Todor Zhivkov became head of the Bulgarian Communist Party (BCP). After ten years of minor changes Zhivkov announced a comprehensive set of economic reforms decentralising economic decision making, restructuring wages to link them more closely to profits and instituting a more flexible pricing system. The reform was halted in 1968 in the face of a severe economic downturn. Throughout the 1960s and 1970s Zhivkov failed to move the country out of its worsening economic crisis. As the domination of the USSR over its satellites began to loosen in the latter half of the 1980s, Bulgaria once again moved in the direction of more comprehensive change (White, Batt and Lewis 1993).

The reforms of the 1980s were meant to devolve power to workers and managers and lead to the production of more marketable goods. The administration of the country was reorganised, and greater authority over prices and wages were given to individual firms. However, in 1988 Zhivkov cancelled the reforms because they were not producing the results he expected. Bulgaria failed to match the economic performance of the West or even other East European states, and the policies of the BCP had produced periodic economic crisis and shortages of goods.

In May 1989 violent protests broke out among Bulgaria's ethnic Turkish population and the government sent in its special forces. At the end of May it was announced that all Turks would be able to leave Bulgaria for Turkey, and some 300,000 Turks fled across the border. The exodus caused immense hardships for an already floundering economy as the Turks constituted a significant proportion of the agricultural workforce. To counter the food crisis the government announced firms would remain open on Saturdays, vacations would be cancelled and youth brigades would serve longer tenures (Europa 1994).

During the summer of 1989 the flight of the Turks continued, the economic crisis sharpened and social turmoil spread. What followed was the reorientation of the Communist party; the elections in 1990, in which the reformed party won a governing majority; and then its abrupt disintegration as a unified and powerful force. Bulgaria faced uncertain political turmoil as its two principle political groupings, the Bulgarian Socialist Party (BSP) and the Union of Democratic Forces (UDF) fragmented into new political parties. The movement towards western democracy is certainly not what the Communist reformers meant for their country in 1989, but it is the goal towards which the country has steadily progressed. However, in similar vein to Romania, considerable economic, social and political change remains to be achieved.

Ex-Czechoslovakia

Roosevelt agreed with Stalin to allow Soviet troops to liberate Czechoslovakia after World War II. When Czechoslovakia emerged from the war Edward Benes again became President and signed a friendship treaty with Stalin. Elections in 1946 gave the Communists a strong position in the National Front, the coalition government of four parties from the Czech territory and two from Slovakia. In 1948 the likelihood of a weaker showing in the forthcoming elections led the Communists to stage a 'coup from within'. Non communist police officers were dismissed and the cabinet resigned in protest. The elections never took place and the Communist Party pushed Benes into suspending the elections and appointing their members to fill the cabinet vacancies. The democratic leadership either fled the country, was forced into exile or was quietly eliminated. Benes died in

September 1948 allowing Klement Gottwald to assume the Presidency.

After seizing power the Communist Party adhered rigidly to the Soviet model. Centralised planning and heavy industries dominated the economy. Large scale industrialisation was brought to Slovakia, significantly less developed than Moravia and Bohemia. The pattern of intensive development continued into the early 1960s when its failures became overwhelmingly clear (Zonis and Semler 1992).

In 1962 reform economists tried to decentralise economic decision making, but fearing that dispersing economic control would lead inevitably to the erosion of Communist power, Antonin Novotny, who had become party secretary following Gottwald's death in 1953, resisted.

Challenges to Novotny's rigid Stalinism increased as economic stagnation became impossible to ignore. In November 1967 students organised demonstrations against poor housing conditions which grew into general political protest. With support from Moscow, the Central Committee replaced Novotny with Alexander Dubecek, the first secretary of the Slovak Communist Party.

In April Dubecek announced an action program which called for the restoration of civil liberties in a democratic socialism. The Kremlin could not accept the Action Program because of its reformulation of the role of the party. When the Czechoslovak Communist Party called a Congress formally to approve the Action Program and to purge the remaining advocates of traditional Communist rule the USSR decided to use force. The Soviet Union with token units from its 'fraternal socialist' allies Poland, East Germany, Hungary and Bulgaria (but not Romania) invaded Czechoslovakia on the night of 20th August 1968, seizing the reform leaders. The Czechoslovakian army was ordered not to resist, although the people resisted courageously and non violently but succeeded only in slowing the reimposition of political orthodoxy.

Dubecek was replaced by Gustav Husak who restored Stalinist controls, obliterating any hint of a civil order while sapping the economy of any remaining vigour. He increased his political strength and in 1970 conducted a party purge. The purge was deep and changed the relationship between the Czech and Slovak republics. He ousted some 500,000 party members, nearly all Czechs, a full third of the party membership. By differentially punishing the Czechs, the Slovak Husak greatly damaged the already strained relationship between the two peoples. In the 20 years following the purge the relationship never improved as Husak continued to favour Slovaks. Disproportionately higher investments were directed to Slovakia and larger numbers of Slovaks were promoted to national office, traditionally filled by the more numerous Czechs. The purge had a debilitating effect on the party and the state as those promoted to replace the purged officials provided an ineffective political and economic bureaucracy (Europa 1994).

The most important event in Husak's 20 year regime was the emergence of the powerful, though small voice of dissent known as Charter 77. After the signing of the Helsinki Accords in 1975, a group of isolated individuals and former cultural leaders began to meet. In January 1977 it released its first charter calling for the government to abide by its own laws and international agreements, particularly the human rights provisions of the Helsinki Accords.

Although the group remained small with at most 2000 members, it became the main force of Czechoslovak political dissent. It stimulated other groups into action and maintained public awareness of issues of human rights and democracy. Charter 77 served as the leading moral force for change in all of Eastern Europe.

In its only significant change of power prior to its fall the party replaced Husak as first secretary with Milos Jakes in December 1987, and Husak was given the Presidency - a figurehead position. Jakes appointed no reformers and introduced no reforms because of profound disagreements within the party on how to maintain its dominance. While Communist parties were falling in neighbouring states, Jakes and his colleagues could only agree on an all or nothing stand. With no capacity for internal reform, the impetus for major change in late 1989 came from outside the Party. Charter 77, the voice of dissent for over 10 years was to form the core of the newly formed democratic government. On December 10th 1989 President Husak swore the new forum dominated government into office and then resigned. The new regime was named the Government of National Understanding and was to stand until free elections could take place in June 1990. Havel, who had taken over the Presidency from Husak, emerged victorious in the country's first post war democratic elections. By June 1991, all Soviet troops had withdrawn from Czechoslovakia and bilateral relations since then have been amicable. More recently, Czechoslovakia has split into two separate states, Slovakia and the Czech Republic. These two have taken quite different paths with the Czech Republic moving towards a freer economy based upon market forces, whilst Slovakia has turned back towards state interference (White, Batt and Lewis 1993).

The maritime sectors

Having established some of the background developments for each country we can now turn our full attention to their respective shipping industries, which have continued to operate and develop throughout the period of political, economic and social chaos outlined in the previous section.

Poland

The modern history of the Polish maritime sector dates back to 1919 following the regaining of the country's independence and with it access to the sea. The government's involvement in the construction of port facilities in Gdynia in 1926 led to the establishment of the Polish Steamship Company (PZM), which still exists today, based in Szczecin. In 1930 the joint stock company Polish

18

Transatlantic Ship Company was established. This company, which engaged in passenger shipping and general liner trade, was nationalised in 1951. It has operated until the present day under the name of Polish Ocean Lines.

The Polish maritime sector and supporting service sectors are now dominated by five operating organisations which have an effective monopoly of business in their fields. Formerly this position had been protected by government regulations and although these have now been swept away, new competitors have only recently begun to emerge and on a small scale as yet. The major organisations are (Ernst and Young 1990):

1   Polish Ocean Lines (POL) based in Gdynia concentrates on the liner trade. Its fleet consisted of 104 vessels at the end of 1989. Of the 5.1 million tonnes of freight carried by POL in 1988, 43% was cross trade and 17% transit. The company made losses in the early 1980s but was believed to be operating more profitably in 1989.

2   Polish Steamship Company (PZM) specialises in tramp operations, and in 1986 operated 149 vessels with a total tonnage of approximately 3.85 million deadweight (Fairplay 1987). The freight carried in 1988 was 24.3 million tonnes of which Polish direct trade contributed 65%. Long term contracts are an established practice in PZM with Polish and large foreign shippers. The financial standing of PZM was sound in 1989 and this had been the case for more than a decade.

Both POL and PZM with average ship ages of around 14 and 11 years respectively in late 1989 have been involved in attempted fleet modernisation programmes to meet the challenges of fierce international competition.

3   Polish Baltic Shipping Company was established in 1976 largely as a specialist ferry operator. It owned 11 ships with a tonnage of 12,400 dwt at the end of 1989. The first ten years of operations were not very successful, but with government subsidies, advantageous credit arrangements and by introducing severe austerity measures the company has survived.

4   Polfracht ship broking and chartering company was established in 1951 as a state owned monopolistic organisation for booking and chartering of cargo in Poland and as the general agent for Polish shippers and Polish shipping companies. In 1988 it lost all legal advantages but its present position in the Polish shipping sector remains very strong and unchallenged. In 1988 the volume of cargo freighted through Polfracht was 28 million tonnes of which 24 million was on Polish vessels.

5   C. Hartwig group of state owned companies are freight forwarders and handle approximately half of Poland's foreign trade (1990).

Other active participants in the maritime sector include two other principle ship owning companies Zegluga Polska Spolka Akcyjna who charter their entire dry bulk fleet to Polish Steamship Company and Polskie Towarzystwo Okretowe whose container and multi purpose vessels are chartered to Polish Ocean Lines.

Another company, Chipolbrok, was established in 1951 to operate liner services between Chinese, Polish and West European Ports. By 1992 the company owned a fleet of 21 ocean going multi-purpose vessels with a total tonnage of approximately 418,000 DWT and a container capacity of over 15,000 TEU's. Meanwhile there are other relatively minor shipping companies such as Transocean, Zegluga Gdanska, and Zegluga Szczecinska.

Following recent changes in the economic policy of the country, the sector is undergoing a gradual transformation with a growing number of newcomers entering such fields as ship chartering, freight forwarding and port services. This can be expected to continue.

Organisation of the maritime shipping sector in Poland has evolved during the postwar period in line with the general direction of the evolution of the economic and socio political system in Poland as a whole. The traditional three stage hierarchical structure of the command type economy, ie the ministry - branch association - enterprise, had already been abandoned in shipping in the late fifties and in other maritime sub sectors by 1982.

The Polish merchant fleet at the end of 1988 consisted of 256 ships (4.1 million dwt), of which 89 were bulk carriers, 46 containerships (including ro-ro and semis), 18 specialist vessels (tankers, ferries etc) with the remainder being relatively small and dry cargo vessels.

The 1988 fleet carried 30.8 million tonnes of which 18.2 million was on direct trades, 10.5 million on cross trades and 2.1 million on transit operations. The total seaborne freight in and out of Polish ports was about 40.1 million tonnes of which seaborne trade with the European Community in 1988 was 13 million tonnes. The Polish - European Community seaborne trade has been dominated by Polish exports (90% of total) of which two thirds were coal deliveries.

The Polish government introduced, at the end of 1989, major new policies directed towards liberalising the economy, introducing as far as possible market principles, privatising the state's assets, stabilising the economy and making the zloty convertible for Polish organisations and individuals. Prior to this the majority of the economy had been managed in a traditional centrally controlled manner although the degree of control had been progressively relaxed in the 1980s in parallel with political developments as noted earlier.

The Polish maritime sector has always been in a privileged position with respect to the degree of autonomy granted by the government administration in respect of operational decisions. The practice of charging freight rates, at least indirectly driven by the international market, was introduced into the industry much earlier than in other sectors, and from at least the 1970s.

However, Poland is undergoing the most radical restructuring ever attempted, and the Polish maritime sector faces a difficult period of adjusting to the new environment (Lloyds Shipping Economist 1990). As an industry which traditionally has operated with a high degree of independence and to a large extent in international markets, this sector is expected to be among the best performers in the new Polish economy.

Successive Polish governments since the economic transformations have pinned their hopes on privatisation of the economy, but while a number of early high profile candidates may be successfully sold off, serious problems exist. Most notably, there is a lack of funds within Poland, but their are also problems of valuing companies in a meaningful way.

Two other main hindrances to economic development are infrastructure and finance; for example the Polish telecommunications systems can take several hours to make an international connection, and there is a lack of fax machines which rely on the same lines. Also, Poland's external debt stood at around $40 billion in 1990 with interest arrears estimated at $7.1 bn and remains a continued drain on convertible currency reserves and the economy generally.

Polish trade declined during the 1980's as economic upheaval followed the unrest in 1980/1981 and martial law. During 1990 a number of influences merged to exacerbate this trend: general economic contraction, falling domestic demand and the gradual break up of the CMEA. By volume, coal exports and oil imports remained the dominant trades. The coal trade is in decline and together with the end of cheap Soviet oil will be detrimental to the Polish trade balance. The decline in Polish sea borne trade has been accompanied by an increase in the reliance of the Polish fleet on cross trades, although domestic trade remains a significant source of employment.

In the new climate POL, PZM, PBSC and other shipping companies will have to choose whether to privatise. They may also need to choose foreign partners and expand into new areas, whilst countering competition in areas where they have been dominant.

Competition has already been experienced in the export of Polish crews, where Morska Agencja w Gdynia (MAG) have been very active. An estimated 6000 Polish seafarers are employed on foreign flag ships and since 1988, 10-20 smaller agencies have emerged to compete with MAG and three other state run companies. Companies such as MAG which are not capital intensive, have few incentives for full scale joint ventures. They would prefer to work with foreign partners on a commission based, project specific basis.

POL believe privatisation will be relatively simple in view of the ease with which assets, if not the whole company, can be valued. For the time being central ship operating activities will remain in state hands; the Treasury funded most of its asset base and it is likely that it will retain an interest in the company even if some form of sell off is embraced. However, plans are advancing to split ancillary service sections of POL into separate companies which may be sold or retained. Operationally it is felt that this will improve services.

21

POL also has a substantial fleet renewal requirement - the average age of the fleet in 1990 was 14.6 years. Polish companies have been free for several years to order overseas if financing was available. One route to western credit under the new Polish legal framework is through a joint venture with a western company. This may involve co-ownership of the entire enterprise; however POL feel that any collaboration should be confined to specific projects.

The other major operator of Polish shipping, PZM, is likewise taking a measured approach to privatisation, with similar plans for the breakaway of units providing support services as subsidiaries or independent companies, while the main company will remain in some sort of partnership with the Treasury. No maritime joint ventures were under consideration up to 1992, although ancillary activities were being developed in conjunction with western financiers, for example a hotel development in Szczecin.

Overviews of the state of Poland's merchant fleet are shown in tables 1, 2 and 3, which provide fleet statistics for the three main shipping companies POL, PZM and Polish Baltic Shipping Company.

Table 1
Polish Ocean Lines' fleet 1986 - 1992.

Type	1986		1988		1990		1992	
	No.	DWT	No.	DWT	No	DWT	No	DWT
DC	43	534,370	40	501,766	40	487,730	11	159,656
DN	42	377,053	37	346,223	33	304,506	2	24,808
WA	16	226,569	15	184,409	11	99,200	6	37,574
DR	11	70,435	4	24,555	0	0	0	0
RF	4	24,114	2	11,314	3	17,701	0	0
PM	3	10,030	2	5,000	3	20,031	2	7,938
JN	2	10,933	1	5,510	0	0	0	0
CL	3	68,400	3	68,400	2	45,600	1	21,305
PU	0	0	1	7,170	0	0	0	0
WL	0	0	2	45,242	2	45,242	1	22,603
Tot	124	1321904	107	1199589	94	1020010	23	273884
Av. Age	12.18		13.46		14.59		8.65	

Key to vessel types - appendix 1. Source: Fairplay World Shipping Directories

Polish Ocean Lines' statistics show a gradual decline in the size of the fleet from 124 vessels in 1986 to 94 vessels in 1990, however, in 1992 the figure is drastically reduced at 23. This is partly due to the initial privatisation processes taking place which have split the company into a number of sectors leaving the core of POL both limited and specialised. The changes are aimed at greater efficiency, and positive results are reflected in the significant reduction in the average age of vessels from 14.59 years in 1990 to 8.65 years in 1992.

One such division has led to the creation of Euroafrica Shipping Lines Company Ltd, which has POL as a parent company. By the end of 1992, Euroafrica was listed as having eight vessels, mainly of DN type, with a total deadweight tonnage of 54,819.

Table 2
Polish Steamship Company fleet 1986 - 1992

Type	1986		1988		1990		1992	
	No.	DWT	No.	DWT	No.	DWT	No.	DWT
BS	67	2291948	70	2580504	55	1972452	48	1744999
DN	41	245,361	40	241,942	33	190,150	2	7,228
BN	27	548,786	3	469,886	25	543,356	56	1069226
TO	5	498,682	3	206,924	3	206,924	1	145,680
DC	2	8,830	2	8,821	2	8,821	2	8,821
TC	2	19,596	2	19,596	2	19,596	2	19,596
TQ	2	19,449	2	19,449	2	19,449	2	19,449
MB	2	198,000	0	0	0	0	0	0
AN	1	15,650	0	0	0	0	0	0
TN	0	0	0	0	0	0	1	1,345
Total	149	3846302	143	3547122	122	2960748	114	3016344
Ave. Age	9.22		10.15		12.92		11.35	

Key to vessel types - see appendix 1
Source: Fairplay World Shipping Directory - various years

PZM statistics show a gradual reduction in the total size of the fleet from 149 vessels in 1986 to 114 vessels in 1992. The 1992 fleet is made up of a significantly higher proportion of BN type vessels, their number at 56 being almost double that of any previous year. These would appear to have replaced DN type vessels which reduced in number from 33 in 1990 to 2 in 1992. The average age of vessels rose from 9.2 in 1986 to a peak of 12.92 in 1990, but fell slightly to 11.35 in 1992.

Table 3
Polish Baltic Shipping fleet 1986 - 1992

Type	1986		1988		1990		1992	
	No.	DWT	No.	DWT	No.	DWT	No.	DWT
DN	6	7,738	?	?	10	14,300	3	5,760
PV	6	6,451	?	?	6	6,592	6	6,495
BN	0	0	?	?	0	0	2	2,141
Total	12	14,189	?	?	16	20,892	11	14,396
Ave. Age		20.42		?		12.44		14.50

Key to vessel types - see appendix 1
Source: Fairplay World Shipping Directory - various years

The fleet of the Polish Baltic Shipping Company is shown in some detail in table 3. Unfortunately there is no available data for 1988 but the overall trends can be gained from the remaining figures. The total number of vessels has changed little over eight years from 12 with a total deadweight tonnage of 14,189 in 1986 to 11 with a deadweight tonnage of 14,396 in 1992; however, the average age has fallen significantly to 14.5 in 1992, from 20.4 in 1986.

Notably, Fairplay lists nine new ship owning companies for 1992, and their details follow:

Przedstowo Polowow Dalekomorskich I Uslug Rybackich Gryf
Own one JN vessel of 1,480 DWT.

Maritime Reefer Transport
Own one RH type vessel of 49,977 DWT.

Morski Institut Rybacki
Own one QF type vessel of 1,200 DWT.

Polish Scandinavian Shipping Lines Ltd
Own one PF type vessel of 2,541 DWT.

Polish Ship Salvage Company
Own one OB type vessel.

Polskie Towarzystow Okretowe SA
Mentioned earlier, Polskie Towarzystow Okretowe SA is a principle ship owning
company whose container and multipurpose vessels are chartered to POL. More
specifically , they own the following:

Table 4
Polskie Towarzystow Okretowe SA fleet 1992

Type	No.	DWT
DC	22	223,467
DN	21	210,919
PM	1	2,500
Totals	44	436,886

Key to ship types - see appendix 1
Source: Fairplay World Shipping Directory 1993

Deep Sea Services & Fish Handling Co. TRANSOCEAN
Own one RH type vessel of 46,387 DWT.

Wyzsza Szkola Morska
Own one JX type vessel of 705 DWT.

Zarzad Portu Szczecin
Own the following:

Table 5
Zarzad Portu Szczecin fleet 1992

Type	No.	DWT
AR	2	679
GT	2	6,990
Total	4	7,669

Key to vessel types - see appendix 1
Source: Fairplay World Shipping Directory 1993

Perhaps one of the most notable changes in the sets of figures is the rise in the registered number of ship owning companies in Poland, from just three in 1990 to twelve in 1992. This increase is most likely a consequence of changing economic and political factors within Poland, encouraging the initial proliferation of new, if small, players in the market.

Romania

In the days prior to Ceausescu's overthrow in late 1989, almost a quarter of the Romanian merchant fleet of around 290 vessels, comprising more than 5m dwt, was laid up because the finance required to carry out repairs to bring them up to class was not available (Lloyds List 01/11/91). Furthermore, the government requirement that its fleet had to serve the domestic economy alone, meant deploying vessels on long ballast voyages to loading ports resulting in a huge drain on resources.

By the summer of 1991 following the demise of the Ceausescu regime in late 1989/ early 1990, and the collapse of Communism's control, a great deal of change had taken place. Captain Marinescu, Romania's Secretary of State for Maritime Affairs, explained 'At this moment we have only eight vessels laid up world wide, three bulkers and five general cargo ships. They are laid up because of sub-standard equipment on board which was manufactured in Romania. We will get these vessels back in service with the help of overseas money and co-operation, and we will strive to ensure our future place as a competitive nation in the international market' (Lloyds List 01/11/91). To obtain this finance the Romanian shipping industry was actively turning to negotiations to set up joint ventures, procuring the help of established western organisations to ensure efficient operation of their fleets, and utilising expertise which for many years was oppressed for political reasons.

The first success came at the end of July 1991 when the Greek company Ermis Corporation agreed jointly to operate five 65,000 dwt Panamax size bulk carriers and two Afframax tankers for Petromin, the organisation set up in Romania in 1990 to operate the bulk carrier and tanker fleet. All seven vessels were taken off the Romanian register and transferred to Malta.

On 12th September 1991 Lloyd's List reported that Economou, the large Greek shipping services group, had established a subsidiary in Romania as part of its plans to expand cautiously into East European markets, initially by creating agency offices. The Piraeus based company, which includes shipping agencies, ship supplies, crewing, repairs and travel among its activities, opened its new office in the port of Constantza, and said it was considering a presence in several other countries in the region (Lloyds List 12/09/91).

By November 1991 Romania was involved in discussions with the Greek seamen's unions with a view to employing Romanian crews on board Greek flag vessels. It was recognised that a rise in unemployment within the Romanian fleet was inevitable, by virtue of the fact that Romanian vessels were overmanned in comparison with western-owned vessels.

In November 1991, talks were also in progress with Piraeus-based Gourdomichalis Maritime SA which would result in the Greek company providing management facilities for more Petromin vessels. Keen to utilise first hand ship management knowledge available in Piraeus, some Petromin personnel were to be transferred to the Gourdomichalis office to help oversee the operation. This move was similar to the management venture with Norwegian owner T Klaveness when a new company Petroklav was set up to manage two Petromin vessels.

Petromin's association with UK based United Dutch shipping was also developed. United Dutch guaranteed around $20m to bring six 65,000 dwt bulkers up to class and operate them in a joint pool (Lloyds List 31/10/91).

In the meantime, Romania has enjoyed a sizeable domestic new building programme. Construction work in the country's three major building yards is expected to last until late 1995 covering construction of bulkers from 25,000 dwt to 165,000 dwt, tankers of 85,000 dwt and general cargo ships from 8,700 dwt to 15,000 dwt.

The Romanians also planned to build ships overseas. In Korea, Hyundai showed interest in building a series of bulkers for the Romanians and although they would need to borrow money for such a venture, Korean banks were actively assessing the possibility of providing the necessary resources. However, the Romanian shipping industry in 1992 were still paying back to the State money which was paid out to build up the fleet during Ceausescu's regime. All those contracts hold 25 year repayment terms.

As to the future, Marinescu concluded that 'the coming years will be difficult. We have to correct all that was wrong but we are convinced that we will become an internationally competitive nation' (Lloyds List 1/11/91).

These changes and developments in western co-operation and to the industry have to be seen in the light of restructuring in the Romanian maritime sector. In July 1990, immediately post revolution, the Romanian fleet of Navrom had split into three parts (Armar: Cpt A Stoica & R Newton MD):

1   Navrom -   general cargo vessels 2500 to 15000 dwt. 1991 figures indicate 90 dry cargo ships (Fairplay 1991).
2   Petromin - tankers and dry bulk carriers. 1991 figures show 80 vessels.
3   Romline - intended to be for specialised vessels but allocated some general cargo vessels to equalise size of classes; it includes for example, container ships, ro-ro vessels and ferries. Totalling 70 vessels in 1991.

The split was carried out to improve control and efficiency of financial and organisational management. Each company has now been split internally into subdivisions for example by size of ship for bulkers; again this is for the same management reasons.

In 1991 the state still owned the larger proportion of the companies - between 75 and 90% - although the aim was to reduce this figure to 51%. It remained unclear who would purchase the remaining parts of the shipping companies. The government is overwhelmed with political and administrative problems and legislative needs and is unlikely to hurry shipping privatisation, although the necessary structures are slowly emerging.

Meanwhile, in market terms, Armar - Romanian shipping agents in the UK - suggested in 1992 that Romanian involvement in the cross trades must have increased to compensate for the drastic fall in exports, due to the change of emphasis from exporting as many goods as possible to earn hard currency pre-revolution, to that of providing for the domestic economy now.

Romanian shipping also has been affected by currency issues and in particular the dramatic change in exchange rates from 14 lei to the £1 in 1989 to around 370 lei in 1991, and up to 900 lei in 1992, making the acquisition of hard currency even more important.

Another major change to have taken place in the structure and organisation of the industry is that all income generated by shipping goes to the shipping companies, who then use this income to pay ship repayments, taxes and disbursements. Prior to 1991, the Romanian government took all shipping income and returned an undisclosed sum of hard currency and Romanian lei reputed to be limited, irrational and inconsistent. However to date no further detailed information is available in this area, for example regarding tax rates. According to Armar the Romanian fleet is 'not highly profitable'.

An overall impression of the state of the Romanian merchant fleet can be gained from table 6. The figures show a gradual decline in the size of the fleet from 298 vessels in 1986 to 255 vessels in 1992, and a notable decline in deadweight tonnage from 5,775,875 to 4,217,972. Also, there is a notable rise in the average age of the fleet from 9.45 years in 1986 to over 13 years in 1992, which places extra pressures upon building programmes. Perhaps the most notable change to have taken place is the creation of new ship owning companies, the majority of these emerging from divisions of Navrom.

Bulgaria

The Bulgarian maritime industry and its ancillary sectors are covered by three state owned enterprises which are still monopolies or virtual monopolies in their areas. Although there are no remaining formal obstacles to overturn this monopoly no strong competitors had emerged by 1993, and none were on the immediate horizon.

These enterprises were outlined by the Commission of the European Communities in 1990 (Ernst and Young 1990) as :

Table 6
Romanian fleet 1986 - 1992

Type	1986		1988		1990		1992	
	No.	DWT	No.	DWT	No.	DWT	No.	DWT
DN	107	617,113	112	687,888	105	676,542	83	605192
DC	73	723,645	72	697,120	72	714,818	67	665,970
BS	44	2281174	46	2316684	45	2384378	30	1502003
TN	10	473,332	11	644,232	6	342,450	8	344,100
TO	8	917,021	8	917,021	8	916,553	4	498,157
BN	19	299,110	19	299,110	18	272,610	33	320,849
WA	14	78,365	14	78,365	11	63,365	9	53,470
BO	5	268,076	5	268,076	6	446,826	4	93,853
PM	1	12,000	2	24,000	1	12,000	2	24,000
CN	2	16,500	2	16,710	2	16,710	2	16,710
ZO	1	0	1	0	1	0	0	0
JN	1	6,000	1	6,000	1	6,000	1	6,000
DK	3	10,109	3	10,291	1	6,777	0	0
RH	10	73,380	10	73,030	10	72,977	10	72.977
BC	0	0	0	0	0	0	1	4,800
TR	0	0	0	0	0	0	1	9,891
Tot	298	5775875	306	5815795	287	5932006	255	4217972
Av. Age	9.45		10.79		10.48		13.17	

Key to vessel types - see appendix 1
Source: Fairplay World Shipping Directory - Various Years

1   'Bulgarian Shipping Company' based in Varna and with the following business activities - maritime transport of cargoes and passengers, handling and forwarding, and ship repair. The company also includes the following subdivisions: the Public Company, Navigation Maritime Bulgare (PC NMB), Port Varna, Port Bourgas, the ship repair plant 'Odessoss' in Varna, the Board of Material and Technical Provision, the Research

Institute of Shipping, and two training centres. The merchant fleet is operated entirely by PC NMB from Varna. Bulgarian Shipping Company is the main liner and tramp carrier in Bulgaria. Established in 1947 it is now one of the largest firms in Bulgaria and one of the major shipowners in the Black Sea region.

2   Bulfracht Co, which is based in Sofia, undertakes the booking and chartering of cargoes and ships. It was established in 1965 and developed as the state organisation for booking and chartering of cargoes and ships in Bulgaria and as the chief agent for the Bulgarian shippers and shipping companies. Under the new economic reforms Bulfracht lost all legal advantages associated with its monopoly in its field of operation. However, being an internationally reputable organisation with highly qualified staff and an extensive network of agencies, Bulfracht remains very strong.

3   Despred Company with offices in Sofia is the main domestic and international forwarder in Bulgaria. The maritime sector is served by two affiliated branches in Bourgas and in Varna. The company was established in 1947. As an international forwarder it handled in 1992 nearly 40% of Bulgarian foreign trade. Until the 1989 economic reforms Despred was a monopoly. Having an extensive network of agencies in the country and abroad, and on the basis of their experience and contacts, Despred retains a strong position in its main activity of domestic and foreign forwarding.

The Bulgarian maritime fleet comprised 120 ocean going vessels in May 1991 (Lloyds List 13/5/91), and as noted earlier, the main owner of the Bulgarian maritime tonnage is Bulgarian Shipping Company. Its ships are operated by PC NMB. The economic circumstances of the seventies and eighties, as well as the financial difficulties of the State during the eighties are heavily reflected in the average age of ships operated by the company. At the end of 1990 the average age of these ships was 14.29 years, rising to 15.38 by 1992.

There are two other small companies in Bulgaria which own and operate ships. One of them is SOMAT which has several ro-ro/passenger vessels. The company operates an intermodal system for transporting cargoes from west Europe to Iran/Iraq (Roe 1991). It uses the crews and the technical services of the Bulgarian Shipping Company. The State Establishment for Maintenance at Sea Channels and Ports' Aquatorium Salvage Tugs (MSCPAST) has some maritime tonnage which it charters out to the Bulgarian Shipping Company. These ships are old but still good for navigation. Bulgarian Shipping Co. provides crews, maintains the ships, and meets the costs.

On 24th October 1991 a decree of the Council of Ministers dissolved the NMB state group, dividing it into a number of sections including ports (Varna and Bourgas), Baltic and Black Sea Shipping Company, NMB (shipping only), and a number of other companies. Assets of the original NMB were divided between these new companies, and the new NMB took over assets and activities of some subsidiary companies, for example the Balkan and Black Sea Shipping Company. Under the new decree, the Balkan and Black Sea remains a stock company with the whole stock to be controlled by the government initially. The share holding structure is now ready for privatisation.

The major problem of privatisation is finance - finding buyers for the companies. As an international activity, shipping is appropriate and perhaps more attractive for privatisation than many other sectors, although due to its international nature it may be argued that shipping is less affected by the country's problems and privatisation may not be necessary. Cpt Domoustchiev - Chairman and Managing Director of the Balkan and Black Sea - felt that privatisation may not be necessary if an alternative method of motivating the industry could be devised, for example profit sharing with employees (Interview 26/11/91). Other problems associated with privatisation in Bulgaria include limited public cash, with only a few potential investors who would dominate subsequent ownership as a result.

To improve image and to satisfy the west, shares might be sold cheaply or even given away, but this would not generate any of the much needed capital. An alternative is to use western involvement. This is seen as important in Bulgarian shipping as it may provide a method of:

- renovating and/or renewing vessels;
- providing know how;
- gaining market penetration;
- improving management skills.

The final path taken may involve a combination of western investment and selling shares to employees and possibly to the public. The Bulgarian state is unable to help with the provision of capital for shipping as it has other more pressing demands for the little capital available.

In 1991 the taxation of shipping companies stood at 40% on profits, 10% on municipal tax, and 2% of other taxes. There was also a tax placed on all salary increases in foreign currency, which represented a 25% tax on profits. Any remaining revenue is kept by NMB. The majority of Bulgarian shipping expenditure is in hard currency and this presents a major problem; working for domestic charterers, they are paid in local currency (leva). The state run bank will exchange leva for dollars, but at a punitive rate and one becoming increasingly so.

According to Domoustchiev, the Bulgarian fleet needs a replacement rate of at least six vessels per annum, as the average vessel age in 1992 reached 15.38 years. The taxation rate precludes this at present, but if tax rates were lowered it might be achievable. More recently the fleet has been pushed towards higher levels of cross trading as the domestic trade falls away and the need to earn hard currency increases.

An overall impression of the Bulgarian fleet can be gained from Table 7. The

31

Table 7
Bulgarian Fleet 1986-1992

Type	1986		1988		1990		1992	
	No.	DWT	No.	DWT	No.	DWT	No.	DWT
DN	45	338,680	39	292,499	37	275,426	25	194,460
BS	21	684,434	21	684,434	21	684,434	21	684,434
TN	10	193,097	9	117,822	7	98,711	6	97,138
WA	3	31,300	1	6,235	3	19,200	3	19,200
TR	6	142,244	6	142,244	4	84,244	0	0
DC	8	98,378	11	119,224	9	102,628	10	116,128
FT	2	24,870	0	0	0	0	0	0
TO	5	460,281	5	280,695	4	183,900	5	280,695
BN	13	218,741	12	222,429	14	290,789	29	397,528
JN	1	6,000	1	5,525	1	5,525	1	5,525
PV	2	15,431	1	4,931	1	4,931	1	4,931
CN	2	18,918	2	18,918	2	18,690	3	27,635
WB	0	0	2	20,868	2	20,868	2	20,868
PM	0	0	2	24,870	2	24,870	2	24,870
TC	0	0	0	0	4	12,800	4	12,800
RF	0	0	0	0	1	1,394	0	0
TR	0	0	0	0	0	0	4	84,244
Tot	118	2232374	112	1940694	112	1828410	116	1970456
Av. Age	10.8		12.49		14.29		15.38	

Key to vessel types - appendix 1. Source: Fairplay World Shipping Directories

figures show a relatively stable fleet size with NMB operating 118 vessels in 1986 and 116 in 1992, and only a modest reduction in deadweight tonnage. However, there is a gradual increase in the variety of ship types - a change which is most likely linked to catering for market demand. Perhaps the most notable figures are those indicating a rise in the average age of vessels from 10.8 years in 1986, to over 15 years in 1992.

In Autumn 1948, the Czechoslovak joint stock company Metrans was founded with a monopoly in the sphere of international forwarding and shipping, including the operating of sea going ships. In 1952, a part of the Metrans' business related to maritime shipments was transferred to the newly established company with monopoly rights - Cechofracht. Its task was on the one hand to arrange forwarding services and, on the other hand, to operate shipping with its own vessels. In 1953, Cechofracht was transformed into a foreign trade enterprise. On a decision of the Czechoslovak government, Czechoslovak Ocean Shipping was established on April 1st 1959, as the first shipping agent in the Czechoslovak history. As an international joint stock company, it was put in charge of operating ocean shipping, purchasing, selling and chartering ships, executing orders of third parties, and taking part in foreign as well as internal trade. Czechoslovak Ocean Shipping took over from Cechofracht all the ships operating under the Czechoslovak flag (Machota 1989).

On 1st January 1991 the fleet consisted of 18 ocean going vessels; of these seven were less than two years old and only four were over 15 years (COS 1990). There were no plans afoot to sell these vessels; a 55% profit tax would make this a pointless exercise. Negotiations are taking place to reduce this figure for ship sales once privatisation has taken place.

In 1990 the company produced its first annual report along western lines. In 1992 figures indicated that the company employed 1200 workers, 100 at central offices and 1100 seamen.

In previous years Czechoslovakian trade dominated the company's trading patterns and the most frequented ports were in Poland, Yugoslavia, East Germany and the Soviet Union, as the ex-Czechoslovakia had no ports of its own. Trade has changed dramatically since then and today the cross trades dominate and a wider variety of ports are used. Figures supporting this were presented by COS during an interview (5/2/92):

| 1988 | 75% of cargo carried (tons) - Czechoslovakian trade |
| | 25% of cargo carried (tons) - cross trading |

| 1991 | 2% of cargo carried (tons) - Czechoslovakian trade |
| | 98% of cargo carried (tons) - cross trading |

This represents a substantial change in 3 years and presents problems of adaptation; however it does mean that convertible currency can now be earned in some quantity whereas previously it was mainly saved by avoiding the use of foreign owned ships.

Under the old system COSCO was obliged to work with Cechofracht (the state forwarder) to carry Czech imports and exports. Today they have complete freedom to act in the best interests of the company. This explains the move toward cross trades, and the need to increase these has been heightened by the decline in the ex-Czechoslovak economy. COSCO still works with Cechofracht, although they no longer have a monopoly on forwarding, and as a state company COSCO are obliged to offer them first refusal on traffic. However, full market rates are still charged.

Although still owned by the state, management of COSCO and its policies are now fully independent, which has been the case since 1989/1990. COSCO is involved with tramping but has no specific trades and no conference membership; their main aim is to keep the fleet fully occupied. As at February 1992, half the fleet was operating on long term time charters, with the other half operating on short terms. Until 1989 some countries such as Taiwan, South Korea, and South Africa were closed to trading due to Czechoslovakian government bans. Recent changes have meant that these are now opened up, except for Israel which remains unserved. Earlier times also saw preferential traffic (eg Cuban sugar) and some forced traffic (eg Chinese - to pay for power stations through barter) being directed through COSCO. Similarly, these requirements no longer exist.

Ninety nine percent of COSCO shares are held by the state owned company FINOP, the remaining 1% belong to Cechofracht which is 100% state owned. Privatisation, as a method of improving efficiency, was planned for June 1992, when it was proposed that FINOP be sold to the Czechoslovakian public through coupons. All individuals have a right to the coupons which can be obtained for a fee of 1000 crowns (then about £20). The fee is to cover administration costs and does not represent the value of the company. Private companies, known as privatisation funds, have been established, which can buy the rights to coupons from individuals not wishing to invest. They can also administer coupons for people by a method similar to unit trust funds. It is expected that 90% of FINOP will be bought by banks and 10% through coupons, although this is still some way off.

A small number of company shares are reserved for employees who do not require coupons. At present the consequential high degree of uncertainty in the management of COSCO, has led to as few decisions as possible being made. Once privatised, COSCO will receive no indirect subsidies and it will have the potential to become bankrupt. Non nationals are unable to purchase shares but when the company is privatised and a stock exchange is opened shares may be bought and sold at will by any person or company.

As to joint ventures the company up to 1992 was not involved in any deals as they were awaiting financial advice and legal agreements on a number of possibilities. Delay was also partly due to Czechoslovakia being split into two separate countries, Slovakia and the Czech Republic.

Convertible currency has been seen as an important issue for some time and remains one today. In the past, the artificial rate of exchange between US$ and crowns meant that hard currency needed to be earned. Now, the rate varies weekly but crowns are freely convertible in both the Czech Republic and Slovakia. The only important factor now is the increasingly harsh exchange rate. In the past, like all East European countries all convertible currency went to the state who gave some back and also paid for new ships. Now COSCO keeps its convertible currency, minus tax, but also has to finance new vessels through profits or mortgages.

Other changes have included taxation rates which are aimed to come into line with western countries. Taxes on profits have altered as follows:

Year	1989	1990	1991/1992
Tax level	80%	61%	55%

The wage bill is also taxed at 20%. Wage levels will remain controlled by the state until privatisation; at present seafarers are paid a mixture of US$ and crowns whilst at sea, but only crowns when ashore.

Further details of the ex-Czechoslovakian fleet are shown in tables 8, 9, and 10. The fleet has remained relatively stable, showing, if anything, a small increase in deadweight tonnage. The most noticeable changes have emerged through the division of Czechoslovakia into Slovakia and the Czech Republic. The Czech Republic retained the majority of Czechoslovak Ocean Shipping which became Czech Ocean Shipping, whilst the Slovaks formed a new company, Slovak Danube Navigation, with just three vessels. The other notable change is the rise in average age of vessels from 6.4 years for Czechoslovakia in 1988 to 8.79 and 9.5 years respectively for the Czech Republic and Slovakia in 1992.

Table 8
Czechoslovak fleet 1986-1990

Vessel type	1986		1988		1990	
	No.	DWT	No.	DWT	No.	DWT
BS	7	241,523	10	434,169	8	339,507
DC	9	112,814	9	106,931	9	107,880
DN	4	23,834	3	17,911	0	0
BN	1	22,623	1	22,623	1	22,623
Totals	21	400,794	23	580,934	18	470,010
Ave. Age	8.14		6.43		6.68	

Key to vessel types - appendix 1. Source: Fairplay World Shipping Directories.

Table 9
Czech fleet 1992

Vessel type	1992	
	No.	DWT
DC	9	108,699
BS	9	311,833
DN	1	22,623
Totals	19	443,155
Average Age	8.79	

Key to vessel types - see appendix 1
Source: Fairplay World Shipping Directory 1993

Table 10
Slovak fleet 1992

Vessel Type	1992	
	No.	DWT
BN	1	3,641
DN	2	4,800
Totals	3	8,441
Average Age	9.5	

Key to vessel types - see appendix 1
Source: Fairplay World Shipping Directory 1993

Summary

As a whole the countries studied show an overall reduction in fleet size and a corresponding reduction in total DWT, although this is sometimes accompanied by an increasingly wide range of vessel types. Generally it might also be said that average age of vessels has increased, and in recent years there has been an increase in the number of registered ship owning companies.

The fundamental social, economic and political changes that have taken place in East Europe have notably affected the shipping industry at every level. It is to a more detailed examination of the relationship between these aspects and an attempt to relate them more specifically that we now turn our attention, with the development of the structure of the research that follows.

3 The research outline

Having looked at the background to East European change and to the East European shipping industries we now need to relate the two. The relationships and reaction of the shipping industry to such change can be viewed in terms of competitive positioning of a service industry (East European shipping) in a context of major macro change. In attempting to analyse these relationships and their effects, we shall divide the task into a series of stages. The aim of the first part of the research in attempting to analyse the effects of recent political, social and economic changes in the Eastern bloc upon the region's shipping industry, represents a broad and complex task which will be facilitated through narrowing the area to a specific geographical region. Poland has been chosen for this purpose because the shipping industry and economy in general are relatively well documented, it is a large maritime nation compared to most other East European countries - owning approximately 37% of the total DWT operated by the five countries selected earlier (ie Poland, Romania, Bulgaria, Czech Republic and Slovakia) - and the political, social and economic situation is not as chaotic as that of the Commonwealth of Independent States, Romania or Bulgaria, where for example, divisions of the ex-Soviet Union are leading to arguments over asset ownership (Lloyds List 02/09/91).

The issues that will emerge are likely to be both numerous and varied. For any meaningful comparison to be carried out the research area needs to be narrowed further to simplify the analysis. The liner shipping market, rather than Polish shipping in general, has been chosen for reasons of data availability, the close contacts maintained with POL, the major liner shipping company within Poland (POL maintains a global network of 22 liner routes and is among the largest 20 container operators in the world), their active involvement in the cross trades, European Community interest and competition on these trades, and the potentially

38

sizeable impact of expanding trade in East Europe upon liner shipping.

Having narrowed the research area we can now turn our attention to specifically modelling the impact of East European changes upon the Polish liner industry. With the increasing power of micro-computers it has become possible to develop models that can assist in judging future trends in the shipping market, such as those summarised by Stopford (1988). One of the most common maritime activities is the preparation of models for forecasting and market research studies, but because the shipping industry is international in its operations and highly complex in its structure, it is often difficult to achieve useful and useable results.

Although models have been widely used within the shipping industry (Dinwoodie 1988, Beenstock and Vergottis 1993), forecasting generally has a poor reputation in maritime circles (Evans & Marlow 1991). The argument that forecasts are never right has been put forward often by shipowners. Beck, planning director at Shell UK adds 'When looking at forecasts made in the 1970s one can find many failures but few successes. Indeed one may be shocked at the extent to which the most important forecasts turned out to be wrong' (Beck 1983).

One example comes from forecasts of demand for new ships produced between 1978 and 1984 which were outlined by Stopford (1988). Each successive forecast predicted a different pattern of demand over the following seven years. The 1980 forecast predicted 50% more demand in 1986 than the 1982 forecast, and even this proved to be too optimistic. In defence of the experts who produced these forecasts there were developments in the world economy that reasonably could not have been anticipated. However the fact remains that they were consistently wide of the mark. Before introducing maritime forecasting Stopford points out that 'the case against forecasting seems formidable' - a view supported by Drucker (1977).

The poor record of maritime forecasting and modelling, and the extensive range of influencing factors present in particular in the East European shipping market have led to the need for an alternative and original approach in the research. This is reemphasised by the fact that the link between East European change and the subsequent impacts upon the shipping industry represents an entirely new and extremely complex situation and no substantial research work has yet been carried out in this area.

These problems and the complexity of this research mean it is not possible to use a conventional econometric model - such as the application of multiple regression techniques - to analyse subsequent development of the shipping industry, which comprises numerous complex social issues and involves almost continual change. A potentially useful alternative has been found in contextual modelling - an adaptation of a specific variant of conceptual modelling - which can be applied to the specific case of Polish maritime activities.

Although few references exist, successful examples can be found in the fields of computer data modelling (The Open University 1989) and environmental appraisal, where in the latter Joyce and Williams (1976) identified several separate contextual sub-models including both operational and theoretical contexts, in analysing the relationship between the urban environment and those affected by it. These models were used because of their ability to accommodate the broad nature of social, political and economic data. Whilst incorporating all data characteristics, they also highlight their inter-connectivity, strength, relevance, and significance to the relationship under study.

The application of contextual modelling is carried out by dividing the broad subject area - in this case changes in East Europe and their subsequent effects on Polish shipping - into a series of contexts, these being constraints upon and issues relevant to the research which help to direct the work and identify the most significant factors which relate to it. The changes taking place in East Europe may be considered as forming the macroenvironmental element of the marketing environment described by Kotler (1993). Macroenvironmental discussions such as those by Kotler (1993), Lancaster and Massingham (1993) and Gross, Banting, Meredith and Ford (1993) have identified several contexts and their sub elements, which have been adapted to apply more specifically to this research.

Once the contextual model has identified the specific relationships between economic, social and political change in East Europe and the maritime sector, what will then be needed is a method of extracting the major issues that emerge and which dominate this relationship, from which it will be possible to begin to assess the adaptation in the market place of the Polish shipping industry to the changes which have taken place. No existing suitable technique is available to achieve this, so the contextual matrix model has been developed by the author, using a variation of a broad strategic matrix approach. The use of matrices as a display technique is supported by Jain, Urban and Stacey (1981), Clark, Gilad, Bisset and Tomlinson (1984), Westman (1985) and Miles and Huberman (1994). Possibly the best known display matrix was developed by Leopold et al (1971) for evaluating environmental impact. The usefulness of this matrix was demonstrated in the assessment of exploratory drilling sites in Antarctica (Parker and Howard 1977), and by the North West Water Authority (1978) when looking at the potential environmental impacts of extracting water from different catchment areas. In Canada, the matrix has been promoted by the Federal Environmental Assessment and Review Office (FEARO) as a screening method to overcome the weaknesses of both project and environmental based screening methods (FEARO 1978). Fortlage (1990) supports their use as they enable all aspects of a problem to be taken into account, but emphasises that they are relatively undeveloped as yet. One relatively recent example of their use is provided by McDonald and Leppard's study 'Marketing by Matrix' (1992).

By setting out and comparing the elements of the contextual model the matrix enables second order causes to be clearly identified. These would, for example, involve the effects of economic change on social issues within the maritime sector. This would include the fact that economic change in Poland has led to suppressed demand resulting in unemployment in the shipping sector. Through the use of consistent terminology, the most important issues will emerge as they will occur repeatedly.

The emerging issues will indicate the main market impacts for Polish liner operations, and demonstrate the need, if it exists, for a repositioning exercise. Positioning and repositioning form an important element of service marketing, to which we will now turn our attention.

In agreement with a number of service marketing texts - such as Cowell (1991), Lovelock (1991), Congram and Freidman (1991) and Morgan (1991) - Kotler (1993) describes a service as

> any activity or benefit that one party can offer to another that is essentially intangible and does not result in the ownership of anything. Its production may or may not be tied to a physical product.

Shipping would appear to fit this description, a view supported by Whiteman (1981) who specified that services include the transport of both goods and passengers by bus, rail, air and sea, but exclude private motoring.

According to Cowell (1991) services have a number of characteristics, the most common of which were identified as the following:

1. Intangibility
Services are essentially intangible in that they cannot be heard, seen, tasted, smelt or felt before they are purchased.

2. Inseparability
Services often cannot be separated from the person of the seller, and creating or performing the service may occur at the same time as full or partial consumption of it. Goods are produced, sold and consumed whereas services are sold, then produced and consumed.

3. Heterogeneity
It is often difficult to achieve standardisation of output when providing services - each 'unit' of a service may differ from other units, and from the customer's view point it is difficult to judge quality in advance of purchase.

4. Perishability
Services are perishable and cannot be stored - for example, empty cargo space on a ship's voyage represents capacity lost forever. With some services, fluctuating demand may aggravate this feature.

5. Ownership
Lack of ownership is a basic difference between a service industry and a product industry because a customer may only have access to, or use of, a facility.

Payment is for the use of, access to or hire of items.

Each of these five characteristics can be applied to the shipping industry. The service provided by shipping companies is intangible, perishable in that it cannot be stored, inseparable from the person of the seller, heterogeneous in its lack of a standard output, and the use of the service does not result in ownership of the vessel.

The Institute of Marketing defines marketing as 'the management process responsible for identifying, anticipating and satisfying customer requirements profitably'. It has been suggested that service dominant organisations are less market orientated than manufacturing firms. Cowell (1991) offers several reasons in support of this argument:

1   The intangible nature of services may cause greater marketing problems than with physical items,
2   Some service businesses are opposed to the idea of marketing,
3   Many service organisations are small and in direct contact with their customers and may not require the same kinds of marketing approaches as medium size or larger organisations,
4   Some service organisations have enjoyed more demand for their services than they could cope with,
5   Ethical constraints may limit marketing in some service areas,
6   Some organisations have enjoyed monopoly powers in their service field,
7   Quality of management is not as good as in service organisations, with top management often not recognising what marketing is or its importance (Stanton 1981),
8   There is relatively little published work in the area of service marketing.

Cowell (1991) suggests that in recent years managers in the service industries have shown greater interest in the relevance and applicability of marketing concepts. He continues by stressing that marketing does have relevance to the service sector but that knowledge of the status of marketing in the service industries is limited, disjointed and under-developed - a concept that is supported by the very limited amount of shipping related marketing literature that exists.

Against this background of service marketing we can now examine positioning in some detail. Trout and Ries (1972) introduced the concept of positioning in the 1970s in a series of three articles describing the origin of the idea. They described the 1950s as the 'product era', the 1960s as the 'image era' and the 1970s as the 'positioning era'. They argued that it is imperative that marketing messages be designed to communicate clearly a product's and/or company's position relative to the competition. By 1986 they claimed that in order to be successful, a company must become competitor oriented and attack weak points in competitors' positioning strategies (Trout and Ries 1986). Green, Tull and Albraum (1988) reinforce the importance of careful positioning.

Organisations that sell to consumer and industrial markets realise that they cannot appeal to all buyers in those markets, or at least to all buyers in the same way. Different companies will be in better positions to serve certain segments in the market, and each company needs to find the parts of the market that it can serve best. Today's companies have moved away from mass marketing and product differentiated marketing toward target marketing.

Target marketing calls for three major steps. The first is market segmentation, dividing a market into distinct groups of buyers. The second is market targeting, evaluating each segment's attractiveness and evaluating which segments to enter. The third step is market positioning.

Once a company has decided which market segments to enter it must decide what positions it wants to occupy within those segments. Market positioning is arranging for a product to occupy a clear, distinctive and desirable place relative to competing products, in the minds of target consumers (Kotler and Armstrong 1992).

Wind and Robinson (1972) stated that product (brand) positioning referred to the place a product occupies in a given market. Conceptually, the origin of the positioning concept can be related to the economists' work on market structure, competitive position of the firm and the concepts of substitution and competition among products. They noted that increasing attention was being given to product image. This suggested a new perspective on product positioning, one which focused on consumers' perceptions concerning the place a product occupies in a given market.

Lovelock (1991) describes positioning as the process of establishing and maintaining a distinctive place in the market for an organisation and/or its individual product offerings. The theory has been used extensively to market tangible products (Bradley 1991), but examples of its use in the service sector - such as Telex and Avis (Keegan 1989) - despite being successful, have been limited. Shostack (1977) promoted the idea that the basis of any service positioning strategy was the service itself, and presented an approach which suggested that within service systems, structural process design could be used to 'engineer' services on a more scientific, rational basis.

In the service sector it is important to note that, because branding is not common, positioning refers to positioning of the entire service organisation and its services (Friedman 1991). In 1992 Lovelock noted that the issues involved in service positioning are numerous and that in a structural sense processes themselves appear to have characteristics that not only affect market position, but also can be managed deliberately and strategically for positioning purposes. By manipulating complexity and divergence, a service marketer can approximate some of the product analysis and design functions that are traditional in product marketing. Moreover, he introduces the use of blueprints to provide a mechanism through which services can be 'engineered' at the drawing board, as well as a tool

43

for identifying gaps, analysing competitors, aiding market research and controlling implementation. The concept of service positioning is further supported by Morgan (1991) and successful examples of positioning strategies in the service and transport industries include those of United Airlines and American Airlines (Lovelock 1992).

To plan a product's position a company first identifies the existing positions of products currently serving its market segments. It then decides what major product attributes customers desire and selects a position on the basis of its ability to satisfy consumer wants better than competitors. A marketing programme then delivers the position to target consumers.

In planning a positioning strategy, a new business or a new service must start from the beginning; however, an established company or service may be faced with a repositioning task if unsatisfied with its current position, if the position has become outdated, or if the position has acquired negative associations - all three of which apply to this research.

According to the AMA (1991), the positioning or repositioning of a service should follow similar procedures:

Step 1   Consumer research - for example in the form of similarity judgements

Step 2   Development of a perceptual map through multidimensional scaling. On a perceptual map the 'psychological' distances between services are reflected on the dimensions consumers deem relevant in evaluating the service.

Step 3   The perceptual map is used to look for gaps or areas where there is open space and possibly a new position.

A product's position is a complex set of perceptions, impressions and feelings that consumers hold for the product compared with competing products. The marketer can pursue several positioning strategies; for example the product may be positioned on any of the following bases:

1   On product attributes;
2   On the needs it fills or benefits it offers;
3   According to usage occasions;
4   According to classes of users;
5   Against a competitor;
6   Away from a competitor;
7   According to different product classes.

These examples show how positioning is being developed beyond its initial focus on consumer perceptions of the market place, to include even additional consideration of competitors' positions and actions - a concept potentially relevant

44

to this research.

Given the basis for positioning products, the choice of which to use depends on a number of characteristics of the firm, product, market, and environmental setting (Wind 1982). These include:

1   The firm's market position.
2   The positioning used by current competitors.
3   The compatibility of the desired positioning with the consumers needs, wants and current perception of the product's positioning versus its competitors, and the given product class.
4   The 'newness' of the considered basis for positioning and its departure from the current practice in the market.
5   The resources available to communicate the positioning effectively and the compatibility of the positioning with the firm's marketing strategy.
6   The firm's desire for an innovative versus similar image, in relation to its competitors.
7   The ability to develop an effective creative execution for the chosen position.
8   The legal environment.

Although there are no such examples in shipping as yet, it is felt that the adaptation and application of a positioning model to this sector provides a useful method of analysis (Ries & Trout 1972, Shostack 1987, Day & Wensley 1988, Green, Tull & Albraum 1988, Hooley and Saunders 1993). In the context of East European change, however, it may be necessary to examine the repositioning that has occurred in the Polish liner market, from a rather different perspective than that put forward by the AMA (1991), outlined earlier, in that service repositioning in East Europe will involve rather more cooperation, direction, and participation in the liner shipping market, than customer perception. This results from a combination of the forces driving company change in East Europe - ie. political, economic and social change - and the specifically regulated environment of the liner industry. This in turn may well require a rather different approach.

Having decided upon a positioning strategy the company can plan the details of the marketing mix - the set of controllable marketing variables that the firm blends in order to produce the response it wants in the target market. In planning to develop a position or to reposition, a company's marketing mix will indicate where strengths lie and on what basis it can sustain a viable position. The marketing mix can provide a method of describing and structuring a company's position.

According to Booms and Bitner (1981) the marketing mix for services consists of seven elements. An adapted version of their idea is shown in figure 1, which is followed by greater detail of each element:

45

Product	Price	Physical Evidence	People
Range	Level	Environment	Personnel:
Quality	Discounts	Furnishings	training
Level	Payment terms	Colour	discretion
Brand name	Customer's	Layout	commitment
Service Line	perceived	Noise level	incentives
Warranty	value	Facilitating goods	appearance
After sales	Differentiation	Tangible clues	interpersonal
service	Quality/price		behaviour
			Attitudes
Place	**Promotion**	**Process**	customers:
Location	Advertising	Procedures	behaviour
Accessibility	Personal selling	Mechanisation	degree of
Distribution	Sales promotion	Employee discretion	involvement
channels	Publicity	Customer involvement	Customer/customer
Distribution	Public relations	Customer direction	contact
coverage		Flow of activities	

Figure 1: The marketing mix for services (Booms & Bitner in Donnelly & George 1981)

The first consideration in the marketing mix will be the *product*. The service product comprises the range, quality and level of services provided, and requires consideration in particular of the use of branding, and for shipping and transit, the after sale service. The product mix of such elements can vary considerably, as may be seen in the comparison of a small shipping company operating ferry services to neighbouring islands and an international shipping company running a wide range of services for a variety of cargoes to numerous worldwide destinations. The product is the core around which all positioning strategies revolve. The company needs to consider whether there is a new benefit it can offer (for example on time service in the transport industry), a new application it can fulfill (for example extending a shipping service to include a new port of call), or a new product classification it can enter (for example extending a transport service to cover door to door). Blueprinting as outlined by Lovelock (1992) may be useful here - it provides a method for rationalising and visualising the service production and delivery process, and provides an opportunity to look for positioning potential.

The second consideration is *price*, which involves levels of charges, discounts, allowances and commissions, terms of payment and credit. Price may also be used to differentiate services and customers' perceptions of value obtained from a service, and the interaction of price and quality are important considerations. Clearly important pricing issues which relate to liner shipping include service contracts, rebates and level of pricing when compared with competitors.

The *place* element relates to the location of the service providers and their accessibility both physically and by other means of communication, as well as the use of technology in delivering services, and the physical evidence that accompanies most services. Linked to this is the issue of distribution channels used and the extent of their coverage, although the concepts of the traditional channel of distribution for goods are not applicable for services. Significant shipping issues here might include ports served, the location of offices, representatives and agents and the trade routes operated.

Physical evidence includes elements such as the physical environment which is fundamental to the service offer (shipping company offices, furnishings, noise), the facilitating goods that allow the service to be provided (eg. ships, cargo handling equipment) and peripheral physical evidence such as tickets and receipts which are of little independent value. Other significant physical evidence will be the vessels and containers themselves, plus related ancillary equipment such as trucks and rail wagons.

Promotion includes methods of communicating with markets through advertising, personal selling activities, sales promotion, direct publicity and indirect forms of communication like public relations. Promotion is a powerful force in the positioning process; for example American Express has repositioned its credit services to women, solely through advertising. However, Lovelock (1991) stressed that positioning is more than advertising or promotion, and must involve each of the 7 P's which make up the marketing mix for services such as liner shipping. Liner operators sometimes actively engage in promotional activities including advertising and sponsorship, and the supply of promotional items. Market research may also be undertaken.

The final two elements of the services mix, which will require careful consideration are the *people* who provide the service and the *process* required to perform the service.

Firstly, 'people' refers to those who perform a production or operational role in the organisation and those involved in selling; the activities of these people can be critical in the selling of the service (Brundage and Marshall 1980). Davidson (1978) suggested that in a service industry the secret of success is recognition that the customer contact personnel are the key people in the organisation. The relationship between customers also needs to be considered, as one customer's perceptions of the service may be influenced by those of another. Clearly significant issues here are the quality of seafarers and officers on board vessels, and the attitude and skills of office personnel ashore.

Secondly, process involves the overall operation of the system including policies and procedures adopted, level of mechanisation, employee discretion, customer involvement, information flows, appointment and waiting systems, and capacity levels. Although traditionally operational management concerns, these aspects affect customer perceptions of the service and are also of concern to marketing management. Performance, people and process are inseparable in many service industries and shipping is no exception.

Each of the 7 P's must be considered when formulating a marketing plan as

47

ignoring any of them could influence its overall success or failure. The specific marketing mix adopted will vary according to circumstances, and the mix must be reshaped in response to changing market needs. There is much overlap and interaction between various components of the mix and decisions cannot be made on one component without considering the effects upon the others. Also the precise nature and impact of elements in the mix and their importance at any one time will vary.

Cowell (1991) also points out that the marketing mix for services should be used with caution because it is intuitively rather than empirically based. Borden (1965) indicates that arriving at the marketing mix is both an art and a science - a theory supported by Cannon (1980). Marketing managers need to undertake research in the markets for which their respective marketing mixes are shaped.

In practice the content of marketing plans varies considerably although a key element in any planning will be the development of an effective marketing strategy. McDonald (1980) supports the theory that this involves two related tasks, namely selecting a target market, then developing a 'marketing mix' for each market.

A major task of marketing management is to blend the elements of the mix so as to meet the needs of each target market. This process must take account of other variables such as resources and objectives of the enterprise, and aspects of the external environment. Formulation of a marketing strategy for services must also take into account some of the unique aspects of services such as :

1   the intangible nature of a service may make consumer choice more difficult
2   inseparability of the producer and the service may localise service marketing and restrict consumer choice; whilst
3   perishability prevents storage of the service and adds risk and uncertainty to service marketing (Bessom and Jackson 1975)

As discussed earlier, a marketing mix can provide a method of structuring an analysis of a company's position. Hence the marketing mix for Polish Ocean Lines requires careful consideration in order to assess any repositioning strategies. When a firm or provider establishes and maintains a distinctive place for itself and its offerings in the market, it is said to be successfully positioned. In the increasingly competitive service sector, effective positioning is one of marketing's most critical tasks.

Positioning is important because it relates to differentiation, commonly recognised as the key point of successful marketing, it forces difficult decisions that focus upon the key areas, it has the potential to make the consumers' selection task easier, and it can help the company to acquire and retain the position of market leader within its industry, or at the other end of the spectrum, to survive substantial external change (Green and Tull 1978).

Positioning methods

So far this chapter has only hinted at the methods which might be used to measure this type of positional change, or might be adapted for such a use. If the repositional activities of POL's liner sector are to be compared with those of the EC competitors, then a 'measurement' method will be needed. These methods fall into two broad groups - quantitative and qualitative - which will now be discussed in greater detail. (It should be noted that the references used in some of the following sections stem in particular from the 1970's, reflecting the limited amount of research carried out in these areas in recent years.)

Quantitative

Quantitative methods may be further divided into the following sections:

1. Spatial representation Methods for spatial representation of structures in data are continuing to appear in ever increasing number and variety and many of these have the potential for application to positional measurement. The confusion that this creates has been aggravated by the widespread introduction of completely different trade names for what are slightly different variants of already existing methods. Papers that report the use of 'proximity analysis', 'smallest space analysis', 'elastic multidimensional scaling', 'Kruscal's M-D-SCAL program', 'Guttman-Lingoe's SSAI', or 'TORSCA' are all referring to basically the same type of analysis - namely the kind of non-metric multidimensional scaling originally introduced and perfected by Shepard (1962a, 1962b) and Kruscal (1964a, 1964b). All of these terms refer to specific methods or particular computer programs that share the following three properties:

i   they are based upon the same basic model (one that assumes a monotonic relation between interpoint distances and the given data);
ii   they use an iterative procedure of adjusting the coordinates for the points to achieve a closer and closer approximation to this desired monotonic relation;
iii   they yield spatial representations that are typically indistinguishable, for practical purposes, when applied to the same matrix of data.

Alternatively, papers that report the use of 'multidimensional scalogram analysis', 'parametric mapping', Kruscal's 'MONANOVA' or Carroll's 'INDSCAL' are referring to quite different types of analyses based upon fundamentally different models concerning the relation between the data and the spatial representation (Shepard et al 1972).
  In 1972 Wind and Robinson outlined the basic concept behind multidimensional scaling techniques. They stated that a market could be perceived as a

49

multidimensional space in which individual brands are positioned. A product's positioning is determined from its position on the relevant dimensions of the similarity space, its position on the various attribute vectors (if a joint space analysis is undertaken), and its position with respect to other brands, as might be obtained through a cluster analysis.

Structural mapping Structural mapping, a derivative of Multi Dimensional Scaling was proposed in a series of papers by Rivett (1977a, 1977b, 1978 and 1980) and Clarke and Rivett (1978), where two methods have been suggested for classifying multi-objective policies. These in turn could be adapted to classify and measure the position of companies within an industrial market, since the original idea of using structural mapping to assess public policies is similar to the concept of commercial service positioning. The first method is to use indifference among policies as input to a multidimensional scaling (MDS) algorithm in order to produce a structural map. The indifference refer to the attractiveness of the consequences, or impacts of the policies, (and potentially therefore, the attractiveness or impact of company service position) and are derived by the analyst's own judgement of the data. No formal arithmetic rules are applied. The indifferences are used as input to the MDS algorithm to generate a map. The map is interpreted to give an ordering of the policies or positions. Multidimensional scaling techniques are concerned primarily with the spatial representation of relationships among behavioural data (Green 1970, Luck, Wales, Taylor and Rubin 1982, Coxon 1982, Davison 1983, Cox and Cox 1994).

Full details of the multidimensional scaling method or structural mapping are given in Rivett's papers, but it is evident that there is a heavy reliance on subjectivity for classifying the policies or attributes of product or service prior to input into the MDS algorithm. Although Rivett argues that this prior intervention is a strength, Massam (1982) believes it a limiting factor. The policy or service attribute indifference table is formulated on the basis of judgement of values in the initial impact matrix. Several analysts are likely to produce different sets of indifference, as might one analyst on different occasions. Also, subjective judgement has to be used to determine the best and worst parts of the multidimensional space in order to produce a ranking of the policies. A multidimension scaling map has no preferred orientation. Different analysts may make different suggestions and a formal criterion is needed to judge each one -a benchmark, or hypothetical best or worst policy, may provide a solution.

Utility values The second method used by Rivett (1977a, 1977b, 1978 and 1980) is an adaptation of structural mapping and employs an additive utility model. In this approach the initial data are converted to a set of personal attribute utility values. These are summed to give each policy or service attribute a final score, and the policies or attributes are ordered on the basis of these final scores. This provides little extra objectivity to the original Rivett methods.

50

Lexicographic ordering This technique can be used to search for a best policy from a set of alternative policies using several criteria (Massam 1980), and can be adapted to provide a ranking of company position on a series of attributes such as price and promotion. It requires criteria to be ordered from the most to the least important, and each alternative policy or position receives a score for each criterion, in terms of its position achieving or moving towards a particular company objective in the market. The scores allow comparisons to be made, and the lexicographic principle of ordering by the first significant difference is used. The alternatives are compared first with respect to the most important criterion. The dominant criterion is declared to be the best. If policies or positions tie for best they are compared using the scores on the second most important criterion. This process is continued down the ranked list of criteria until only one alternative remains. This is the best as it has received the maximum number of highest scores on the criteria.

Difficulties with lexicographic ordering include the lack of information about the magnitude of the difference between criteria. Also there is no account of the magnitude of the difference between the scores for policies or attributes on each criterion. There is no way in the method to trade off scores among criteria, which is perhaps the greatest weakness of lexicographic ordering as a method for determining the best policy.

The requirement for ordering the criteria is valid but by making it a necessary precondition to the analysis many types of data cannot be used. A method has been devised which attempts to get around this requisite. Using every possible ordering of criteria means that overall the criteria will be considered equal and there is no dominant order. This allows policies or positions which are dominant on several criteria to be identified.

Overall, lexicographic ordering does not produce a complete ranking of all the alternatives and does not indicate the degree of difference between the tabled ranks which represent the closest approximation to a summary of relative attractiveness. However it may be used as a guideline to indicate the most desirable position for a company where a set of alternatives is presented.

2. Multivariate analyses Multivariate analyses with potential application to positioning fall into two main types, as follow:

Factor analysis Although a very old and well tested technique, Stewart (1981) pioneered the use of this approach for tackling multi-criteria policy choice problems, such as that presented by the introduction of a positioning strategy. The positional policy choice problem is similar to the traditional classification problems addressed by factor analysis, in that each alternative policy can be regarded as a case, with the impact values as the observations on each variable. The variables are the criteria used for evaluating the positional policies. The purpose of this approach is to reduce a complex array of data to a format that will help identify the relative attractiveness of alternative positional policies. Factor Analysis could be used in positional assessment (and therefore measurement) by creating a series of factors from the criteria that are used to indicate position (such as price and promotion) which are composite and will thus overcome the problem of comparing and combining criteria with different scales, metrics or no metric at all. Its drawback in this research, other than the extensive debate that has surrounded its use and validity over the years (Nijkamp 1979, Stewart 1981), is that it requires large quantities of data on each criterion (for example, sizeable sets of prices over many companies; or prices of many products for one company), to produce statistically reliable information. The context and structure of this research cannot fulfill this requirement as it is examining positioning of a limited number of products (for example 20', 40', reefer containers with a limited number of market segmented products) in a single market. Thus, despite potential it cannot be used. Its application has most potential where the data set is one of customer opinion on company product, and not the product itself.

Concordance analysis   This is a formal arithmetic procedure by which multicriteria policies can be classified, and thus product or service provision, and the results are used as input for a multidimensional scaling algorithm. This could be used to indicate favourable positions and their related strategies. The general procedure is similar to that of Rivett's structural mapping approach, except that the policies are classified objectively rather than subjectively. For reasons stated earlier this is seen as a preferable methodology. Concordance analysis is a technique whereby all possible pairs of policies are compared with regard to their impact for each criterion, and a concordance index is calculated for every pairwise comparison.

The concordance indices are summarised on a concordance matrix, which will be square with the policies on position on each axis. This matrix can be interpreted to determine the relative attractiveness of the policies by calculating a row sum scale, but it is more accurate to transform the matrix into a dissimilarity matrix which can then be input into a multidimensional scaling algorithm. The use of a benchmark is helpful in classifying the alternatives in the multidimensional space. Problems again emerge in this research context because of the type of data required by the technique and that available.

Evaluation of the advantages and disadvantages of the various techniques are uncommon, although McAllister (1980) looked at the overall framework of the appraisal process (and its application to positional problems) and produced a checklist of criteria by which to judge alternative methods. This list was studied by Massam and Askew (1982) who made the following comments:

> Factor analysis appears to be an unsatisfactory procedure, as not all the available information is used - much is discarded to simplify the analysis. Stewart (1981) shows that factor analysis can break down on problems

52

where there are fewer alternatives than criteria - as would be the case in this research.

Of the two methods proposed by Rivett, the use of indifference and utility values have drawbacks as follows:- Firstly, in both cases, the use of a subjective interpretation of the data prior to an objective analytic technique would seem to undermine the validity of using the technique (MDS algorithm and summation). Secondly, both methods assume that the analyst can perceive correctly the marginal rates of substitution between the criteria. For the indifference method this would be intuitive, but the utility value method needs commensurability between the scales by which the criteria are measured to allow the summation to be possible. A usual additive utility model achieves this by data normalisation (Rivett 1980), but the subjective approach is dependent again on the analyst's intuition.

Together these two methods violate two principles of decision making procedures. Being subjective they are unsystematic and non-replaceable; further, the nature of the procedures make the data analyst also a decision maker as the two roles are necessarily inseparable here.

None of the problems discussed here appear to apply to the concordance analysis method. It is objective, permits incommensurable scales, and utilises all the given data. It allows sensitivity analyses to be easily included through varying different aspects of the procedure and presents the set of best positional policies in a way that indicates the relative attractiveness of all the policies. Further, disaggregation of the data to indicate the distributional characteristics of the impacts can be readily incorporated into the procedure. The ability of a procedure to compare policies, or positions, according to the individual groups or sectors of the population affected, is an important quality. However, it remains inappropriate due to the statistical requirements of the technique compared with the availability of data.

Further problems associated with quantitative techniques which could be applied to positioning, include the much discussed areas of measurement and valuation:

Measurement Massam (1980) argues that while judgement, opinion and intuition may be called upon to defend a choice, a stronger case can be made if indisputable measurements are available to show the advantages of a particular selection. He points out that not all individuals have the same point of view and make the same kinds of trade-offs among the various elements involved in comparing alternatives. Measurement is not always possible as there may be indirect effects which require consideration but may not be easily identifiable. Similarly, there may be some effects which are difficult to measure in a meaningful way, for example marketing strategies, workforce attitudes and quality of service.

Valuation Once each aspect has been measured, the positional measurement must be converted into a common metic. This presents further problems, for example, how to commonly value time, risk, pollution, views, attitudes, prices and so on.

The common element of all the methods reviewed so far is that they aim to present a set of better positional policies from a larger set of feasible alternatives. As such the methods may be seen as sieving devices for reducing the set of feasible alternatives, rather than as a means of identifying the best position. The most appropriate method needs to be comprehensive , systematic, simple and quick, and be an aid in the decision making process. Massam and Askew suggested that concordance analysis and multidimensional scaling meet these criteria, but remain fraught with difficulties.

Wind and Robinson (1972) agreed that MDS techniques were especially suited to the portraying of the perceived position of a product. Moreover, they stated that nonmetric MDS scaling and clustering techniques utilise data that need only be rank ordered which facilitates the respondent's task and places the burden of analysis on the researcher.

In discussing the concept and measurement of positioning they described a number of studies in which multidimensional scaling procedures were used to determine the product's marketing position. The use of MDS techniques to measure positioning is further supported by Green and McMennamin (1973), Kotler (1971) and Tull and Hawkins (1980). However, there are a number of problems associated with the techniques, such as the lack of any satisfactory statistical methods available to determine the dimensionality of MDS solutions (Jain, Pinson and Ratchford 1982). More specifically in the case of this research - the study of POL's position - the technique is notably unsuitable for the following reasons:

i   Consumer preferences and perceptions are of limited relevance, as the market is often regulated, being controlled by agreements such as those on the Trans Atlantic and by state intervention.

ii   Differences in the carriers are unclear to many consumers as the market is fixed, and there is often no direct contact between POL or other operators and the consumer, as agents are used.

iii   The existence of fixed market shares, fixed prices, agreements and contracts prohibits the use of MDS.

iv   The changes taking place throughout East Europe, which include political, social and economic issues, are driving POL to reposition its product rather than it being market led by consumers. This suggests that we need to apply a service marketing approach within the context of these changes.

However, Massam and Askew emphasised that each of the methods outlined should only be used as aids in the choice of an optimal positional policy.

Meanwhile, all the techniques suffer from measurement and valuation problems which stem directly from their quantitative ambitions. Hence the qualitative approach.

Qualitative

Qualitative techniques are analytical, non quantitative methods for examining project proposals or company strategies. In terms of assessing position and positional strategy for a company in the market place, most qualitative approaches are derived from project and plan evaluation methods originating in the public sector where the difficulties of aggregating disparate criteria are substantial. Similar problems are met in measuring company positional strategy and hence there are opportunities to use the techniques in attempting to assess the changes occurring at Polish Ocean Lines. The position evaluation or policy choice problem may be simply stated. Given a set of alternative positions, evaluation and assessment are concerned with the selection of one of the alternatives that best fulfils the objectives of the company. The traditional approach to this type of evaluation is the cost-benefit method as used by Pearce and Nash (1981), Walshe and Daffern (1990), and Layard and Glaister (1994). This could be adapted to assess positional strategies of a company to give a valuation of each. Cost-benefit evaluation involves a tabulation of all the benefits and costs of strategy or project, followed by a comparison of the sum total of all the benefits and costs, in order to arrive at an estimate of value. Although in a theoretical sense the cost-benefit method would seem to provide a potential solution to positional measurement, in practice it suffers from the following difficulties:

1   It is difficult to evaluate intangible criteria (for example, corporate image). The estimation of the intangible benefits can often distort the whole value estimation process.
2   The expense of obtaining all of the benefit and cost data needed to apply the benefit and cost method is often prohibitive.
3   Cost-benefit analysis is not easily related to indirect costs of related projects or programmes.
4   It is difficult to allow for various uncertainties of implementation in cost-benefit analysis.
5   Problems of measurement and valuation (including pricing of intangible items) are predominant, as noted earlier.

If another method is to represent an improvement over cost benefit analysis it must overcome to some degree the difficulties listed. What is needed is a method that would handle intangible benefits and indirect costs with less data collection, analysis effort and spurious objectivity (Schlager 1968).

Planning balance sheet approach The planning balance sheet (PBS) attempts to overcome some of the weaknesses of the cost benefit analysis method. The PBS stresses the importance of recording all impacts, whether monetary or not, and analysing the distribution of impacts. Although originally designed for town planning issues, it is appropriate for adapting to company positional analysis as it attempts to measure and assess the variables that describe a multi facetted situation.

PBS is a method for conducting systematic evaluations devised by Lichfield (1960, 1964, 1966, 1969, 1970, 1975) and applied by him on several occasions to town and regional plans in Great Britain. In PBS, as in CBA, impacts are measured in non technical terms and estimated where possible by scientific methods. However in contrast to CBA the evaluator is given wide latitude in making judgemental estimates because PBS seeks to determine impacts at a more detailed level, which therefore is less amenable to scientific estimation procedures.

Impacts are converted into monetary figures whenever possible, but other impacts are also recorded. This means that an overall aggregated assessment cannot be calculated. Instead, the user is required to determine the alternative most advantageous to the planner or company's interest for each impact category. Thus the rating method yields a set of simple sub-indices: one for each impact category (McAllister 1980). In positional terms it could be used to assess the numerous and diverse criteria that reflect a company's approach to the market, whilst attempting to measure or value them all.

Matrices Matrix evaluation techniques are the most formalised of the alternatives to cost-benefit analysis. Numerous derivations exist each of which possess two major characteristics:

1   The type and size of the impacts/policy, are listed in disaggregated form and occasionally also in terms of individuals or groups upon whom the impacts fall.
2   The impacts are then weighted with relative values so that they can be compared to one another and some matrix of performance can be derived. Alternatives are selected on the basis of this index.

Matrix construction is carried out by placing alternatives in the horizontal axis, and objectives in the vertical axis. The measures of how the alternatives perform in relation to the objectives are placed in the cells. This simple process can be elaborated or simplified in a variety of ways to suit particular circumstances. A number of different evaluation matrix techniques have been proposed and these have been detailed by Lichfield (1970) as:

Bending on This is a process developed in the UK whereby the ideal form is agreed upon and then alternative plans are tested to see how well they conform.

Often social and economic objectives are formulated then tested. The best is then selected and 'bent on' to the actual situation in question having due regard to the opportunities and constraints.

There exists only a tenuous link between objectives and the ideal form and there is no certainty that the 'bent on' version is better or worse than the one rejected. Costs are ignored, producers and consumers are not distinguished, nor are the costs falling on them. There is little reference to incidence and benefits are not accurately measured. Although potentially applicable to positioning the failure of the technique to be comprehensive, accurate or specific in its recommendations makes it inappropriate.

Policy Evaluation Matrix This technique was devised by Kreditor (1967), and compares more directly the objectives for the actual plan or company strategy proposals. Alternatives are ranked against a series of objectives to show whether each is affected directly, marginally, negatively or not at all. Altogether this gives a qualitative visual ranking. This is useful but there is no explicit distinction made between costs and benefits, and it is not comprehensive in its consideration of objectives, nor of incidence.

Goals Achievement Matrix (GAM) This is the most well known and sophisticated matrix approach and was developed by Hill (1960, 1968, 1973). It resembles the Policy Evaluation Matrix of Kreditor but the whole issue of objectives is more carefully considered. It was devised as a response to the deficiencies Hill saw in Planning Balance Sheet Analysis and Cost Benefit Analysis.

Kucharzyk (1977) outlined the main characteristics of GAM as follows:

1   All objectives are specified in advance of both the design of alternative plans and the analysis of their consequences;
2   The objectives involved are multi-dimensional;
3   GAM is designed to compare mutually exclusive plans or company positions;
4   The selected objectives are assigned weights to reflect relative importance, prior to actual evaluation. Information for these weights may have to be obtained from attitudinal and behavioral studies;
5   Alternatives are measured in terms of the extent to which they allow progress towards or away from each objective;
6   There is no conversion of scores to a common monetary base. This problem is shifted to objective weighting and index choice;
7   It allows for the effects on different groups to be evaluated and distributional issues can be covered.

In effect, Hill's GAM displays a substantial quantity of information, but it can also be used to decide between alternatives. Three major types of method exist:

1   Goals achievement account - Here there is no synthesis of information at all. The decision about the best alternative is left to the decision maker entirely.

2   Weighted index of Goals Achievement - the combined weight of the objectives and their incidence is assigned to the measures of achievement of the objectives. The weighted indices of goals achievement are then summed and the preferred alternative is that with the largest index. This provides a good summary of information and a comparison of alternatives is easily carried out; however it does tend to hide much that might be useful. Also there is doubt whether such disparate factors can be combined.

3   Goals achievement transformation functions - Ackoff (1962) has proposed a method for relating outcomes measured in different units to avoid some problems confronted when aggregating measures using different units of measurement. However it is a method with many practical problems.

Hill (1972) discussed the advantages he felt would arise from adoption of the GAM which are summarised as follows:

1   It expresses the complexity of the consequences, and the use of all available information enables more rational decisions to be made.

2   It is an improvement in many ways upon PBSA or cost-benefit analysis techniques.

3   GAM includes both qualitative and quantitative aspects.

4   It can be usefully applied and compared with standards, and the costs of reaching such standards can be determined.

5   The weighting of alternatives and objectives gives the method inherent flexibility.

Hill also pointed out the disadvantages and problems associated with applying GAM which can be summarised as:

1   It cannot determine whether a project should be executed or not.

2   The GAM is designed for plans in a single functional sector, and remains of limited application as a result, in the evaluation of programmes which include recommendations for projects in several sectors.

3   The paradigm of rational planning upon which GAM is based, is rarely found.

4   Problems exist in the simultaneous analysis of costs and benefits relating to qualitative and quantitative objects since these are frequently measured on differing types of scales - ordinal, interval and ratio.

5   It is difficult to determine weights objectively, and this is the key to decision making in the GAM.

6   There is the problem of obtaining sufficient reliable data to make the method workable.

Many of these problems are also suffered by the majority of techniques which could be applied to positional assessment, and overall, are not solely confined to GAM but to matrix evaluation techniques in general (Roe 1980). Any form of matrix technique makes simplifying assumptions and the results cannot be treated as sacrosanct. They produce no more than best estimates - although this must not detract from the need to be rigorous in their development. Chmara and Langley (1973) concluded that the matrix is a useful conceptual framework within which a detailed assessment methodology can be developed. The important issue is how the matrix is used and how scales, scores and weighting can be selected for incorporation into it. If the limitations associated with the selection of these factors are recognised, then the matrix approach provides a useful tool in evaluating a project, proposed investment, or positional strategy. It is flexible, allows political weighting to be introduced, can accommodate any number of factors and can be varied in form so that any particular consequences (eg. cost, safety) can be emphasised. Finally its emphasis on goals means that the achievements of any strategy are closely related to the original aims of the decision-makers, and this in turn, helps to ensure a more rational decision-making process.

Frameworks The use of the framework - essentially a derivative of a matrix approach - in plan and project appraisal has been demonstrated by the UK Department of Transport (1977) who concluded that whilst current methods of scheme appraisal based on cost benefit analysis were sound as far as they went, they believed the assessment to be unbalanced and suggested a shift of emphasis in the whole approach. They stated that it was unsatisfactory that the assessment should be so dominated by those factors which are susceptible to valuation in money terms, and they believed it to be inadequate to rely simply on a checklist to comprehend environmental factors. Positioning assessment faces the same sort of multi facetted problems and hence frameworks offer a potential solution.

The Department believed that the right approach is through a comprehensive framework, relying on judgement, which embraces all the factors involved in scheme assessment; also, that such a framework should be employed from the earliest planning stages of a scheme. Extensive examples of framework usage can be found in the October 1979 report of the Standing Advisory Committee on Trunk Road Assessment (SACTRA 1979).

As a display method, a framework is a matrix structure which lists elements on the vertical axis, with scenarios on the horizontal. Information is then sorted into the appropriate section of the framework, enabling all results to be displayed.

In a marketing context, Wind (1982) further supports the use of display techniques such as the framework to measure position, where consumer evaluations of a set of brands are unavailable or inappropriate, as is the case in the context of this research.

The inherent problems involved with quantitative techniques, combined with a basic lack of specific, quantifiable data available to the research being undertaken, have led to the rejection of these methods for repositioning in favour of qualitative techniques, as supported by Bagozzi (1994). Of these, the most favourable is the framework analysis which, although not faultless, has been successfully used in previous transport planning situations (Department of Transport 1977,1979). The use of display techniques is supported by Voodg (1983) who highlighted Chadwick's (1971) criticism of 'reductio ad absurdum' of the qualities of an alternative into one single number, and stated that it is necessary that important distinctions and differences are kept intact in a presentation. As well as the multicriteria planning carried out by Voodg, display techniques have been used widely in Environmental Impact Analysis - for example relating to major European transport projects in the European Community - where the state of knowledge and the techniques will often only permit a qualitative assessment (Department of the Environment (1976) and Clarke and Herington (1988)). Friend and Hickling (1987) provide numerous further examples of the use of display techniques in the area of strategic planning.

The qualitative techniques are analytically descriptive, but the framework approach allows all types of available data to be taken into account and enables comparisons to be made between company positions whilst not excluding quantitative information. It avoids the pitfalls of the quantitative approaches, with their rather spurious objectivity, in attempting to measure and aggregate positional criteria, and provides a qualitative approach that retains detail whilst facilitates comparison and analysis. Quantitative measurement is both impossible and impractical for this research, since we are assessing a wide range of multivariate company activities, not those of consumer opinion, so that only relative movements and positions can be shown in this final stage. It is suggested that the key discussion areas for research into the repositioning of POL are isolated by the 7 P's, which provide structure to the research. Lovelock endorsed this theory in 1984, by stating that positioning, or repositioning, involved each element of the marketing mix. In chapter six, a framework will be used to provide a model for Polish Ocean Lines and its market positional change from 1988 to 1992.

Information for the frameworks was gathered during a number of personal, or individual, interviews as described by Green, Tull and Albraum (1988) and Bagozzi (1994). According to Payne (1982) such interviews can provide useful information for product positioning decisions. The interviews were carefully structured for consistent research (Patton 1987, Adams and Schvaneveldt 1991, Babbie 1992). Prior to each interview an interview schedule was prepared (Baker 1994). Similar to a questionnaire, an interview schedule is a list of questions to be asked, prepared in such a manner that the questions are asked in exactly the same way of every respondent. Examples of the question areas set out prior to

60

each interview can be seen in appendix 2. This form of research is often referred to as standardised interviewing (McNeill 1990).

A potential problem area identified in chapter one was that of language barriers. It was suggested that research into the area of East Europe might be hampered to some extent due to communication problems, particularly relevant when contemplating interviews as a method of collecting information, but also potentially of significance when reviewing published and unpublished literature. However, no such problems were encountered. All respondents were found to speak fluent English due to the fact that this is shipping's international language, whilst the large majority of literature was similarly in the English language.

After the interviews the discourse was analysed, and the responses were applied to the framework (Bryman and Burgess 1994). Although, as with all qualitative methods, this may give rise to some bias, this has been acknowledged in the research, and where possible accommodated - unlike quantitative methods it is transparent in its assumptions.

Having achieved a framework analysis for POL the technique will be repeated for its competitors who are based in the European Community. This will then enable some form of positioning comparison to be made. The comparison will be carried out using a further variation of the matrix approach which has been developed by the author as no suitable technique exists. There will be no attempt to quantify any positional movement, rather a display and interpretation of the results. The use of qualitative approaches, and particularly matrices is supported throughout this thesis, and hence maintains consistency.

For reasons discussed earlier, and again to maintain consistency throughout the research, we will use the seven P's of the marketing mix to provide structure to the comparison. The seven P's for POL will form one axis on the two dimensional matrix, whilst the seven P's for the European Community competitors form the other. Having input information for POL in 1988 and the end of 1992, and the position of European Community competitors for the later date, we can then look at the comparison within each cell in turn until every combination has been achieved. For reasons of simplification and comparison a single general position will be provided to represent the majority position of all of the European Community competitors. This will enable us to display the results and will help to indicate the position of POL relative to the European Community competitors as well as showing any movements which have taken place by using arrows. Major anomalies in European Community position will be discussed in the text.

The complete research outline can be seen in figure 2, showing the models to be used at each stage. Having discussed the methodology for this research we can now look to operationalising each area, which will begin in the following chapter with the contextual model.

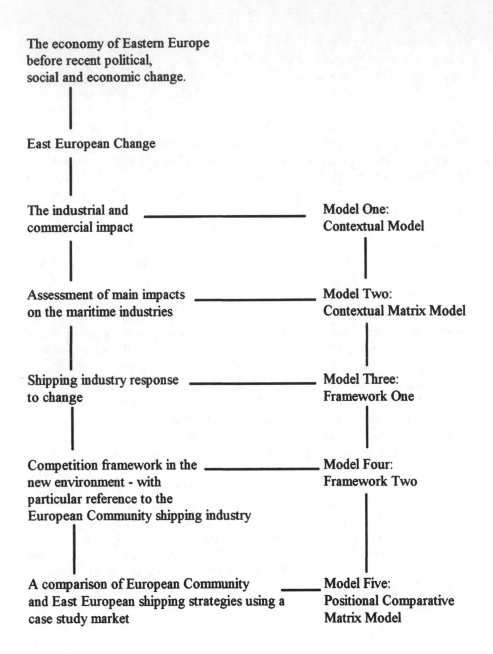

The economy of Eastern Europe
before recent political,
social and economic change.

East European Change

The industrial and   Model One:
commercial impact   Contextual Model

Assessment of main impacts   Model Two:
on the maritime industries   Contextual Matrix Model

Shipping industry response   Model Three:
to change   Framework One

Competition framework in the   Model Four:
new environment - with   Framework Two
particular reference to the
European Community shipping industry

A comparison of European Community   Model Five:
and East European shipping strategies using a   Positional Comparative
case study market   Matrix Model

Figure 2: The research outline - showing model structure and inter-relationships

4 The contextual model

In this chapter we will develop the contextual model to help identify the main issues affecting Polish shipping. Before this, we will examine the economic background to the Polish situation. A broad analysis of Polish economic developments formed part of chapter two, but it is helpful here to examine the country's recent economic trends in a little further detail. The following figures of key economic indicators provide an essential background to the maritime developments which have occurred since the mid to late 1980s.

1982	1987	1988	1989	1990	1991	1992
84.82	265.08	430.55	1439.18	9500.0	10576.0	15500.0

Exchange rate / US$ (average zloties)
Source: Economist Intelligence Unit (EIU) April 1992

These figures represent a dramatic fall in the value of the zloty and mean that foreign goods have become increasingly expensive and Polish demand for them has fallen. The resulting decline in imports and hopefully, rise in exports, has had a consequent impact upon the shipping industry, amongst a number of other effects.

1985	1986	1987	1988	1989	1990	1991
29.7	33.5	39.2	39.2	38.9	46.6	46.5

Gross Hard Currency Debt ($bn)
Source: Economist Intelligence Unit (EIU) 1992

1982	1987	1988	1989	1990	1991	1992
25,900	42,620	42,146	43,029	49,386	38,604	40,554

Total foreign indebtedness ($mn)
Source EIU April 1992

These two sets of figures show a considerable overall rise in Poland's debt between 1982 and 1992. This has placed pressure upon the economy, meaning that finance is no longer readily available for public expenditure - for example on improvements to transport infrastructure (including vessel renewal) which are needed by the shipping industry to increase competitiveness.

1989	1990	1991	1992
1 -	6.6	11.4	12 +

Unemployment (%)
NB: - less than
        + more than
Sources: GUS, Warsawa 1990, FT 3/2/92, Guardian 6/2/92, Ind 23/4/92, ESCEC 14/5/90

The rising unemployment figures have both positive and negative effects upon the shipping industry. Firstly, higher levels of unemployment mean there is greater competition for jobs. Companies have a wider choice of employees and workers are usually willing to accept lower wages. On the negative side rising unemployment and lower wages mean that the demand for products falls, and with it the derived demand for shipping.

Poland	GB	USA	Czechoslovakia	Hungary
1.35	10.56	13.90	2.05	1.70

Average cost of one man hour in 1990 ($)
Source: ESCEC 14/5/90

In April 1992 the average Polish wage was officially recorded as being just below $180 per month (Independent 23/04/92).

The figures indicate low wage rates in relation to other countries, especially those in the west. This again has twofold implications for the shipping industry. Firstly low wages mean low demand for products, particularly those which are imported and therefore comparatively expensive. This in turn means low demand for domestic shipping. Alternatively, the shipping companies need only pay their employees low wages, which keeps costs down and may provide some competitive advantage in the international market.

1982	1983	1984	1985	1986	1987	1988	1989	1990	1991
105	21	15	15	18	25	61	244	585	60

Prices (percentage change on previous year - rounded to nearest figure)
Sources: GUS Statistical yearbook, various years. EIU 1990

1989	1990	1991	1992
700 - 1000	180	60.4	45

Inflation (%)
Sources : Independent 6/3/92, Financial Times 6/4/92, ESCEC 14/5/90

The price and inflation figures indicate substantial price rises throughout the 1980s and into the 1990s, with at times evidence of hyper-inflation. These mean a reduction in the demand for products and a consequent fall in demand for shipping and an erosion of standards of living through a fall in the value of savings and real wages.

Output by industry, 1991 (% change on previous year)
Private sector   25.4
State sector   -24.1

These figures for 1991 indicate a substantial growth in the private sector which is newly emerging since the introduction of privatisation laws. This would produce positive benefits for shipping as it likely represents a rise in exports. However, the fall in state sector production may mean a fall in exports. It is possible here that the figures may be partly due to the privatisation of some state industries so that the net change in production is less significant.

1987	1988	1989	1990	1991
2.0	4.1	0.2	-12.0	-2.0

Real GDP - year on year percentage changes
Source: EIU 1990

The GDP figures denote year on year percentage changes in the market value of the Polish economy's domestically produced goods and services (Barro 1990), and represent in this case a decline in the economy (and associated standards of living) in Poland particularly during 1990. Most recent indications are of a slight rise in standards during 1992, the first in East Europe since the changes.

This selective economic data is not intended to produce a full picture but helps to provide a broad background against which other developments and particularly those in the maritime sector can be analysed.

The methodology of contextual modelling was described in chapter three, and it is to the operationalisation of this model that we now turn our attention. For the purpose of this research the following contextual submodels can be identified as areas of change in East Europe which will have impacts upon the shipping industries:

1   Political - examines the political changes taking place both in and outside Poland and their subsequent effects upon the shipping industry.

2   Economic - looks at new methods of business, material prosperity and general economic changes taking place in Poland and the Polish shipping industry.

3   Spatial - concerns the geography of Poland and its hinterland and its influence upon the operation of the shipping industry.

4   Technical - related to Poland's industrial condition and the effects upon and condition of the shipping industry.

5   Managerial - examines levels of and attitudes towards management skills available in Poland, which affect the Polish shipping industry.

6   Social - looks at the social changes in Poland, the effects of the recent political changes upon the Polish people, and subsequent social and economic effects upon the shipping industry.

7   Legal - concerns changes in Polish law, which will have a direct effect upon the shipping industry, as well as changes in international maritime and non maritime laws.

8   Organisational - related to changes within Polish institutions and their relationship with shipping.

9   Environmental - examines the increasing pressure for shipping to take account of its environmental impact, particularly in the context of rising East European awareness.

10   Logistical - concerns the requirement of shipping to adapt to changes in production techniques and delivery requirements, as part of a complete through service.

Each of the above contexts is inter-related to varying degrees, and the relationships between the various contexts must be logical and pragmatic in order to present a realistic analysis. In order to reveal and understand those interrelationships and to identify the most important issues and constraints it is necessary first to examine each context in greater detail.

Political context

The political situation in Poland is changing, with most movements towards a form of democracy recognised in Western Europe (The Economist 1/2/92). Consequently, this implies that those involved in the shipping industry cannot be ignored by government, as they not only make up part of the electorate, but also take decisions that affect the electorate both directly and indirectly. Perhaps more importantly, democracy involves a transfer of decision making to place authority with individuals, groups and shipping companies, which will call for fundamental changes in day to day operations.

The other political impacts of shipping such as in employment, and satisfying industrial requirements will be key issues when considering future changes. Shipping is not only a major employer but also serves a multitude of other industries which will be affected by any changes to the shipping industry.

In the political context shipping is also in competition with numerous other sectors of the economy. Shipping is not always seen as a key issue and often has a low public profile. It may suffer a lack of recognition and attention, as the government is more likely to want to be seen concentrating on issues that are more popular with or precious to the electorate, such as health care, education and housing. Hence shipping may suffer from neglect in terms of investment, and this combined with shipping's diminishing role in earning hard currency and enforcing state security will reduce the significance of the sector even further.

The removal of security/spying issues relating to shipping is an important political issue. The European Community believed that Eastern bloc shipping had a common purpose of serving its own markets and that it performed dubious defence exercises (Bergstrand and Doganis 1987). All vessels had a defence bias or alternative defence purpose, but with the removal of Communist state control this is no longer the case, and shipping is now able to concentrate on commercial activities.

External political influences such as those from the European Community, Group of Seven (G7), European Bank for Reconstruction and Development (EBRD), and the US, and national rejection of the ex-USSR, are encouraging Poland to follow a free market model which has both direct and indirect effects upon shipping, - one such example involves the US and the International Monetary Fund (IMF) financial plans and encouragement to devalue the zloty.

Political constituents of a new government and the relationship of shipping to the government will also have an impact upon the shipping industry. In 1991 Poland became divided between a multiplicity of small parties with little prospect of forming a stable coalition able to sustain an agreed programme (Financial Times 28/10/91 & 29/10/91), a problem that has continued through 1992. Changes therefore will be slow and chaotic affecting the shipping industry's rate and effectiveness of transition. For example to date there have been slow changes in laws relating to subsidy, prices, bankruptcy and privatisation as a direct consequence of the election results.

The political decision of state withdrawal from many sectors including shipping both financially and from administration is intended to lead to increased competition between individual shipping companies and it is expected that a wide divergence in performance standards and services will emerge. (Lloyds List 11/10/91).

The need to relate market demands to shipping needs is a problem previously dealt with by the central planning organisation of the CMEA. Political withdrawal from the market means that in future this will become a commercial issue, and shipping companies must learn to meet and react to market demands to survive in a competitive environment.

Other political changes relate to the formation of joint ventures. Examples of these from East Europe include deals between SeaLand and ex-Soviet transportation authorities (Fairplay 31/5/90), Petromin and United Dutch Shipping Co (Lloyds List 31/10/91) as well as western ventures with Romline (Fairplay 12/7/90) - although there are few from Poland itself. There are a number of reasons behind the formation of a joint venture; Western firms may look to the East to reduce costs, enter new business areas, share or obtain new expertise, acquire means of distribution, and penetrate domestic markets; East European objectives are diverse but include obtaining new technology, import substitution, access to modern production and management techniques, earning hard currency through exports and helping satisfy home demand (BOTB 1989, Seatrade Business Review 1989, Lloyds Shipping Economist 1990). Also joint ventures are seen as a good move politically both in the east and west, and the European Community and East European governments have encouraged their creation. In October 1991 Lloyds List reported that setting up a joint venture has been one of the only ways for cash starved Eastern bloc shipping companies to slow down the shrinking in the size of their fleets (Lloyds List 11/10/91). However as well as both sides seeing it as on one hand, a way forward, on the other it may be viewed as a 'selling off of companies to the west' for little in return.

Privatisation is considered in the organisational context (see below), but may also be used as a political tool. Privatisation is seen as being popular with the electorate and hence is publicly supported by the Polish government. It is however, argued that privatisation in the shipping industry is unlikely to occur for some time. Polish people do not have enough money to buy into such a capital expensive industry, and if foreign companies were allowed to buy, it is unlikely that they would want to, due to the lack of profits being generated in an industry where at present losses are being made. Despite this, Warsaw has included Polish Ocean Lines (POL), Polska Zegluga Morska (PZM),the ports of Gdansk, Gdynia and Szczecin, ship repair yards and fishing companies, in a list of 400 companies throughout Poland which will be privatised as a priority (Fairplay 15/8/91). When, or if privatisation does occur in the maritime sector it will have direct effects at every level.

Organisational context

One of the major organisational changes faced by the shipping industry involves the introduction of privatisation. Privatisation has implications for every level of operation. For example it may mean higher levels of company investment and increased competition leading to greater efficiency.

Internal shipping company structure is related to competition and co-ordination between departments, which are affected directly and indirectly by economic, social and administrative changes. Recent East European reforms will move towards eliminating cross subsidy between departments in state firms so that each must operate independently and thus efficiently. Free market, privately owned companies will have to adopt similar, Western style structures in order to survive.

The organisational context also looks at the internal organisation of bodies related to shipping; these include shippers and industry, government ministries, freight forwarders, local authorities, unions and the labour force, ports, transport companies as well as banks and financial institutions. Changes to these organisations will have an effect upon the organisation of the shipping industry at the company and broader industrial levels. Any such changes will be related to issues such as privatisation, mergers, competition, and reorganisation along free market lines, each of which will have a direct organisational effect.

Shipping must fit into the free market organisational model proposed for and imposed upon Poland in general by both outside and inside forces. Temporal issues must also be considered, which include the need for short term adaptation, the need for long term major structural change, political timescales and elections, social timescales concerning for example price inflation and consumer and electorate patience, shipping company timescales relating to profits and bankruptcy, as well as international debt, all of which have organisational implications within the firm.

Economic context

Recent changes in Eastern Europe generally, and Poland in particular, have involved economic restructuring, with the introduction of fundamental changes to economic policy.

The introduction of free market pricing policies will affect the maritime industries, as they will encourage a competitive environment for shipping and other sections of the economy, which should enhance efficiency and may lead to cost cutting exercises as seen in the west, as well as the possibility of bankruptcy. The reduced costs likely to emerge due to competition are unlikely to outweigh cost increases caused by inflation as prices rise to a true level (for example in the CIS and Poland between 1990 and 1992).

Inflation as an economic issue is likely to have an overall negative effect on shipping. Increasing costs for example caused by reduced subsidy, the use of hard currency to pay for imports, and raised interest rates, will lead to price increases and a fall in disposable incomes. Subsequently demand will fall, followed by a fall in demand for shipping, due to its derived nature. Inflation has an indirect effect on shipping activities, as well as a direct impact in raising costs such as labour, fuel and maintenance. The cost increases will again mean that there is less capital available for investment.

Another economic policy involves the removal of subsidies from many industrial sectors. This will place new economic pressures upon shipping companies - for example they will have to finance new vessels from their own revenue, especially once privatised, and they will therefore need to be both competitive and profitable. Similarly the removal of indirect subsidies such as those related to building and power provision will push up costs. Also the removal of subsidies from goods will mean an increase in prices for items used by the shipping industry (Ernst and Young 1990).

The economic issue of hard, convertible currency also needs to be considered. Shipping is a hard currency earner at present; it is used to buy foreign goods, which cannot be paid for using the soft currencies of the Eastern bloc. However the Polish zloty is moving towards convertibility and it has been suggested that it will become a fully convertible, hard currency by 1995 (ESCEC 1990) although most commentators think it unlikely to occur this quickly. If this is achieved foreign goods could be purchased using zlotys and businesses will be able to charge for Polish goods in zlotys. This will have two main impacts upon shipping - first it will no longer be interested in earning dollars at any cost by undercutting western competition, and second, prices will rise due to the zloty floating on the international exchange rate market; this in turn will lead to falling demand for goods and hence for shipping. Such a level of convertibility in other East European countries seems unlikely in the foreseeable future, although some progress may be seen to be made in Hungary or the Czech Republic.

Another economic change has been the slow introduction of privatisation, which looks likely to continue. Shipping could be affected relatively early because it may be one of the easier areas to privatise. Reasons for this include its attractiveness to westerners as it involves large capital assets with markets attached; also shipping is in less financial difficulty than many other industries. However, privatisation of the shipping industry may not occur in the immediate future for the reasons mentioned earlier - that it is a capital intensive industry and there is little money available within Poland to invest. However, it is believed that privatised shipping companies would be more willing to invest in new technology needed to raise efficiency and enhance competitiveness (Lloyds Shipping Economist 1990). Examples of privatisation schemes in the maritime and related industries within Poland so far include deals at the Gdansk ship yard (Fairplay 10/5/90), Corona Ferry Lines, Euro-Africa (Szczecin) and some road haulage operations (eg. Pekaes, Warsaw).

Shipping is believed to be a major potential area for joint ventures and will be restructured to attract western investment. Polish objectives in joint ventures have been outlined above and should lead to improved shipping standards (BOTB 1989).

Competition externally on cross-trades provides an economic challenge. Having to operate on the same principles as western companies for the first time means that Polish shipping companies will find it harder to compete on the cross trades. They will no longer be able to operate at less than cost with prices subsidised by the state, as it was widely believed they did under the old regime, but will have to operate commercially gaining profits through more efficient practices and lower labour costs.

Economic competition with other sectors for resources causes further problems. Shipping is increasingly likely to face competition for resources from other sectors of the economy such as manufacturing and energy. This has not been experienced in the past as the direction of resources was centrally controlled, artificial in its execution, and benefitted shipping which was seen as a profitable hard currency earning venture.

Oil resources from the ex-Soviet Union have caused economic problems for Polish shipping. Since December 1989 the CIS has been delivering only 80% of the oil it has contracted to supply to Poland because of production difficulties (ESCEC 1990 Poland). Trade relations between the CIS and Poland need to be reorganised, but the two countries are not prepared for the process, and the political and economic situation in the CIS is too chaotic at present. Initially the reorganisation will have an adverse effect on the Polish economy. Both the problems of oil supply and any reorganisation of trade would affect shipping both directly and indirectly through pushing up prices of oil and derived commodities. Also, oil must now be paid for in hard currency which will create a further drain on the already low resources.

Changes in economic policy are also directly affecting coal and other exports. Producers can now export in whatever way they wish. However, the Polish authorities have imposed a provisional limit on 22 agricultural and industrial commodities to ensure supplies to the domestic market during the transition period (ESCEC 1990). This has a direct effect on shipping companies who will now be dealing with individual producers as well as the many trading agencies; also, limitations imposed upon exports will have a direct impact upon the level of demand for shipping.

Another major economic issue is that of international debt and repayments. In 1990 the Polish economy provided the security for the national debt of $4 bn and roubles 6 bn (ESCEC 1990). The level of repayment that these debts require will mean little finance is available for programmes such as port or road development. This will affect the development of the Polish economy and the speed at which 'westernisation' and improvements to the shipping industry can take place. In parallel, Poland's trade balance with the west has been in the black since 1982. It is possible that this may deteriorate as the low level of competition in the Polish economy is gradually opened up (ESCEC 1990).

New economic demands and practices of banks and financial institutions, for example the application of true interest rates, tighter lending regulations and reduced availability of finance for loans will all have direct and perhaps negative affects upon the shipping industry.

On the positive side, the introduction of free market principles should mean companies becoming free to operate commercially, which may improve the economic condition of Polish industry. For example businesses will be able to retain profits and use them for modernisation programmes and to enhance competitiveness. Also there will be opportunities to gain expertise and finance from the west, which should benefit shipping in the long run.

Economically, the general shipping market should become more open, more level and more competitive as Polish shipping companies adopt western principles including free trade and free pricing.

Technical context

Within the technical context the state of Poland's fleets and ports in particular need to be considered. In 1990 the average age of Polish vessels was fairly high at around 12 years, and by 1992 had risen to over 15 years; meanwhile the state of the ports was relatively poor (Ernst and Young 1990). However, allowing companies to retain their own profits, together with the establishment of joint ventures, may enable greater levels of renewal and general improvements to take place in the long run. Privatisation means that efficiency must rise, whilst joint ventures and retention of company profits should mean that money and expertise are available for improvements to occur. Port improvements would benefit shipping; for example increasing efficiency and decreasing turnaround times will offer greater scope for profits to be made.

The state of ship and port equipment must also be considered. Again new investment opportunities may eventually lead to improvements being made to existing equipment. This should increase the level of shipping efficiency and again could lead to increases in profit levels.

Another technical consideration is the standard of inland transport infrastructure which by the early 1990s was poor with a long history of low investment and little government finance available to make improvements. This caused problems for shipping by inhibiting the movement of goods to and from the ports leading to increased transport costs by delaying vessels, interrupting traffic flow, and contracts being lost. On one hand, new economic policies may lead to more finance being available to improve the situation. However, on the other, a reduction in State subsidy will mean that less money is immediately available and there will be a time lag before any improvements can be made.

On a similar theme, the technical condition of most Polish industries is outdated by western standards but if this improves through joint ventures, privatisation and western involvement, it could lead to increased productivity which will subsequently benefit shipping. However, this is likely to be preceded by a downturn as state subsidies are withdrawn and finances dwindle.

Information technology availability - both soft and hardware - and the extent of related knowledge was low in 1991 compared to more developed countries but is likely to improve with western intervention and expertise, which will raise the level of efficiency in shipping. Again, this is likely to take some time to achieve due to the withdrawal of subsidies in the state sector, lowering the amount of finance immediately available for improvements and awaiting an economic upturn accompanying growth in the private sector.

Likewise, telecommunications are poor at present; for example with very poor standards for national telephone calls, and few fax machines. The introduction of Western expertise and finance, for example through joint ventures, may improve services which in turn would greatly benefit shipping amongst other industrial and commercial sectors.

Managerial context

The recent political and economic changes which have taken place in Eastern Europe have necessitated dramatic changes throughout the workforce, and especially in the decision making role of management. Managerial skills presently available in Poland were not designed to cope with a free market economy. Management needs to acquire western style skills, which may be facilitated through a number of joint ventures, or through training programmes largely provided by the West. The new style economy places new demands on the entire workforce and this requires a change in education as well as training programmes, many of which - for example marketing training - are already available.

Marketing awareness, skills, and its related infrastructure, in Poland and by Polish companies internationally, require some development. Although production was centrally dictated, now that this is no longer the situation, the Poles will need to market their products and services in a more competitive fashion. Marketing skills are most likely to improve with expertise coming from the west, perhaps in the form of joint ventures. In order to undertake these joint ventures new Polish skills and alterations to present methods of organisation are needed in all industries. Indeed, one of the reasons quoted by East Europe companies in support of joint ventures, was to gain some experience and understanding of western management styles.

The attitudes towards management and organisation, and the basic decision making methods need to change considerably in order to cope with and facilitate the more western style businesses that will develop. Ideally there should be a move away from the present downward form of decision direction - where top management push decisions toward the shop floor and customers - toward an upward decision making process, as often occurs in western companies.

Already it has been recognised that basic organisational structures within the shipping industry have had to change to suit new methods of operation. Polish industry and employees are used to central control, the majority of which has been removed by the demise of the CMEA and the decline of state influence. Some companies may need to decentralise further to be able to compete with western companies.

Organisational structures outside the shipping industry, for example in government and in the wider transport sector are also changing as the whole political and social situation changes. These structural changes will have a direct effect on maritime businesses. As an example, levels of government involvement and interference have altered. With the adoption of new economic policies, the Polish government is becoming less and less involved with industries which are increasingly left to their own devices. This has a direct effect and requires a new style of industrial management. Similarly, the adoption of new economic policies has meant diminishing interference in shipping companies from regional organisations, and local interference is now largely obsolete.

Management will also need to adapt to be able to collaborate with related organisations such as shippers and freight forwarders and to communicate with them in a competitive environment. In previous years, central control meant that this was not necessary. The removal of state monopoly privileges has also undermined many established operations. Ship brokers and agents are finding new companies setting up overnight in direct competition. Only a few will survive, but as a market economy all will be given increasing chances to try (Fairplay 15/8/91).

In order to negotiate foreign business deals and to compete on a world wide scale, shipping companies must gain some understanding of how other countries and competitors carried out their business. Such understanding was not previously necessary and again this will require the development of entrepreneurial skills.

The role of unions and labour force have changed in recent years, and they are now allowed greater freedom and power which they are beginning to exercise. Workers are exhibiting more militant tendencies and are becoming less willing to accept decisions outright. As mentioned previously, this is increasing pressure for a reorientation of the decision making direction which will strengthen worker participation and contribution to company management. It is interesting to note that this was the ideal of the old regime, but not often the reality.

Another issue to be considered within the managerial context is the rising level of unemployment. As this occurs labour availability is increasing which may lead to lower employment costs for shipping companies. However, the quality and skills of labour available, particularly at the management and entrepreneurial levels may not match with those in demand; this relates back to the basic issues of management culture, education and training.

A separate issue concerns vessel registration. Solidarity (the Polish trade union) is against Polish owners flagging out their vessels, arguing that protection against poor technical and working conditions would be lost. The government is sympathetic to this, but puts the case that Polish owners needing to approach foreign banks to win credit are often required to reflag their vessels to countries such as Cyprus (Fairplay 15/8/91).

Social context

The reforms taking place in East Europe will have numerous social consequences. Basic issues such as democracy, together with freedom of speech and movement need to be considered. The transfer of the role of decision making to place authority with individuals and companies will have a large social impact which will affect the shipping industry both directly and indirectly. For example, Polish workers will need to acquire new skills in order to cope with a free market economy rather than central control, which has a social as well as a managerial impact.

The level of acceptability of change is also an important social aspect. So far the Poles have suffered the problems of transition with 'remarkable stoicism' (Financial Times 20/11/90). However, if this did not continue, social unrest might lead to strikes or other action which may interfere with the transition of the shipping industry. It may also lead to changes back toward a more traditional Communist style government through the electoral process, as a reaction to the hardships - with clear social implications.

The eventual success of the reform programme depends on the support of the unions and the labour force, but social repercussions such as mass redundancies, tougher discipline and limitation of union rights will be difficult to accommodate (ESCEC 1990) and may cause problems or setbacks for the shipping companies.

Commercialism in Poland has led to severe and novel social problems such as unemployment, which in 1991 was expected to rise to 2 million or 14% of the workforce. Increasing unemployment figures will mean less money is available to purchase goods and subsequently there will be less demand. Ultimately this could lead to a fall in demand for shipping. Perhaps less important is the fact that unemployment leads to lower wages and could therefore benefit shipping by lowering costs, although lower wages will again mean that demand for goods and hence shipping falls even further. Hence, unemployment will have indirect and most likely overall negative affects upon the shipping industry.

Aggravating unemployment problems is the fact that free market pricing policies have led to large increases in the price of both Polish and foreign goods. This again is likely to lead to a fall in demand for goods and subsequently for shipping. In 1989 inflation reached an annual rate of 700 -1000%, and led to a drastic erosion in real incomes. This not only led to falling demand but also meant a subsequent rise in domestic costs of the industry. Inflation is likely to continue as a problem as it was still relatively high at approximately 40% per annum in 1992 and still far from stable.

At the same time shipping faces pressure from other domestic sectors and from abroad. With reference to broader social competition with other sectors, shipping is likely to suffer in that it has difficulty in competing with issues such as health, housing and the environment. European Community and other external demands - for example encouraging democratic progress and full free market pricing - will have direct and indirect affects on the shipping industry in social terms.

Legal context

The present legal framework in Poland is changing constantly, as a consequence of the legal inadequacies of the old regime and the changing social and economic situation. This means that shipping companies need to keep up to date and adapt to many new requirements. Reliable sources of information concerning new laws and regulations must be available to participants in market activity, as well as systems for efficient registry of businesses, property and legal encumbrances. This need is increased by the fact that further changes are likely as other economic, political and social reforms are carried out.

Reform proposals call for the creation of equality of economic opportunities in an environment in which market forces are able to operate. Perhaps the most important step that could be taken in regard to the equalisation of conditions for business activity would be the passage of a clear and detailed law concerning the relationship between the state and its employees and agencies and any business, enterprise, or commercial organisation in which the state has an ownership interest. Such a law could clearly define the permissible and impermissible relationships between state-owned entities and state bodies, providing a management structure with specific safeguards against interference in the financial or business affairs of the enterprise (OECD 1990).

Laws are undergoing constant changes which will affect shipping directly. Recent legislation introduced in Poland has enabled privatisation, and other new business activities such as joint ventures to be carried out (BOTB 1989). Also, the creation of bankruptcy laws has meant that theoretically, bankruptcy of both state and private companies is now possible. Each new law places increased pressure upon shipping companies and often requires a fundamental change in operations and management techniques, particularly following privatisation and the removal of subsidies.

The accounting system also needs to be revised. Accurate and timely information is an essential input into economic decision making in a market economy. The system must provide the information necessary to present a full and fair view of the financial position of an enterprise to allow management to evaluate performance and take key decisions. Audited financial accounts, incorporating a proper valuation of assets and liabilities, will need to be published for the information of shareholders, creditors, and those entrusted with supervisory authority. This will require a number of relatively straight forward changes to the present accounting regulations. Getting managers to apply the new concepts may take time and effort, and it will be necessary to train accountants and auditors as well as financial commentators (OECD 1990).

Problems have arisen in the fact that recent Polish governments have been very fragmented which creates difficulty when attempting to develop and improve laws. Also, the current political chaos makes enforcement of laws more difficult for all industrial sectors.

On a wider scale international law is gaining greater recognition and true acceptability, rather than token acceptance under the old regime, partly due to pressure from organisations such as IMF (International Monetary Fund) and BIMCO (Baltic and International Maritime Conference). This will have direct and important effects upon the shipping industry, impacting upon both operations and practices.

Spatial context

The spatial context concerns the variation in economic activity and the population distribution of Poland, which although mainly fixed, especially in the short term, will help to identify some of the main constraints within the model, and may be subject to some eventual modifications due to the changes taking place in Eastern Europe.

One of the main spatial constraints upon Polish shipping is its proximity or otherwise, to the major international shipping routes. Poland's only access to the sea is along the north coast, at the Baltic, which presents some problems when compared with the more accessible competing ports, of Bremen and Hamburg for example. It is along Poland's Baltic coast that the main Polish ports of Szczecin, Gdansk and Gdynia are all located. Gdansk is located on Dead Vistula, a tributary of the river Vistula; Gdynia is approximately 32 km north west of this, in the Gulf of Gdansk; Szczecin is situated on the west bank of the river Odra 61 km from the open sea. Port location will be a key factor affecting shipping movements, and clearly has a notable impact on the routes chosen for the import and export, of commodities, especially following the removal of controls by the state on port usage.

Within this section the location of internationally competing ports serving East European markets for example Trieste, the Baltic states, Hamburg, and Rostock, will each have an influence upon Polish shipping activities, the ports chosen for operation and the competitiveness of the Polish maritime industry (for example the move of the majority of Polish Ocean Lines' vessels to a base in Hamburg from Gdynia).

Another spatial consideration is the population of Poland, which stands at approximately 37.7 million, of which 60.8% live in urban areas. Poland is divided into 49 provinces and 2365 rural communities which contain about 40,000 villages (BOTB 1988/89). Principal cities are:

> Warsaw - population approximately 1.7 million, the capital and an
> important centre for light industry.
> Lodz - (847 400), textile centre.
> Krakow - (744 000), scientific industry, railways and steel.
> Wroclaw - (640 000), engineering, railways and electrical industries.
> Poznan - (578 000), engineering and railways, trade fairs.
> Gdansk - (468 400), shipbuilding industry.
> Szczecin - (395 000), shipbuilding industry.
> Katowice - (367 300), heavy industrial centre, coal, steel and chemicals.
> Bydgoszcz - (369 500), light engineering.

The location of industry and centres of population will have direct effects upon shipping operations, for example by dictating the type and amount of labour available in various regions, and by controlling the amount of produce for export and the number and type of imports required. Also the changes taking place in Eastern Europe generally are likely to have some impact upon these industries and may lead to movements towards new locations for industry or for the population in general - for example increased migration towards the towns. Any such impacts would have consequent, if indirect, effects for the shipping industry.

Within the spatial context, the location and economic and political progress of transit areas, for example Hungary and the Czech and Slovak Republics, will also place constraints upon Polish shipping activities. It is also important to note here that Poland occupies a key position between East and West Europe (Financial Times 20/11/90), and hence is likely to play an increasingly important transit role as the economies of the region - especially that of the ex-Soviet Union - develop or decline.

Finally, it is necessary to consider the supply of other modes of transport and the extent and standard of competition to the maritime sector, for example the move towards private trucking and away from state directed shipping. Similarly the infrastructure provision for these modes may act as a constraint. This was generally agreed as poor by the early 1990s, and acts as an inhibitor to shipping development. However it is felt that developments in the new economic policy may mean finance is available to improve the situation to the eventual benefit of shipping.

Logistical context

The logistical context refers to the role of shipping within the overall flow of goods from production to consumption. As East European shipping operations come more into line with those of western companies they will have to learn to react quickly to modern developments. The recent emphasis upon improved production techniques such as Just-in-time (JIT), means that shipping companies face changes in delivery requirements for raw and part manufactured materials. Shipping companies need to liaise with manufacturing and warehousing nationally and internationally, and they need to integrate to become part of newly required door-to-door services. These problems will become notably apparent as state direction of logistics and intermodal operations is gradually removed.

New logistical requirements will affect shipping companies' ports of call, and feeders and demand improved reliability. Other aspects which need to be considered here include fleet decline and its impact upon the ability of Polish companies to provide an efficient and flexible service, and over tonnaging internationally.

Environmental context

The changes taking place in Eastern Europe generally, require most industries to develop a greater awareness of environmental issues. Within this context environmental pressures come from funding sources, such as the World Bank, the Group of Seven (G7) and foreign companies thinking of investment or joint venture activities who are concerned about their international image. There is also pressure from the European Community who insist that their legislation - including that concerning the environment - is adopted before Poland can obtain associate or full membership. Further pressure comes from within Poland, which has seen a substantial growth of the green movement in recent years.

In shipping, environmental pressures relate particularly to fuels, coatings, cleaning of holds and dumping of garbage overboard. However, shipping is well used to operating in an international environment where international regulations apply - although the old regime sometimes failed to comply fully with international legal standards.

Overall discussion

Within each of the above contexts, it should be noted that the majority of issues involved occur at different levels, these being international, national and regional. Although some of these have been noted individually, for example the sources of pressure for environmental awareness, they relate to many issues. This activity at differing levels may result in contradictions and conflicts in pressure which present problems for Polish shipping in terms of adapting to change.

For the purposes of the research each of the ten contexts is related to a greater or lesser extent. The contextual model provides the mechanism through which an understanding of these interrelationships may be achieved. The model places constraints upon the research development in that it helps to define its boundaries, as well as helping to identify the major issues involved. However, it is suggested that the overlaps can be drawn out more clearly by developing the model further, as we shall see in the next chapter. This in turn will help to reveal the more significant issues that are directing Polish shipping developments.

5 The contextual matrix model

In the previous chapter, we used the contextual model to identify the specific relationships between economic, social and political change in East Europe and the maritime sector. We will now operationalise the Contextual Matrix Model, discussed in chapter three, in order to extract the major issues that emerge and which dominate this relationship, from which it will be possible to begin to assess the adaptation in the market place of the Polish shipping industry to the changes which have taken place. This will be done using the concept of second order causes, assessing the inter-relationships of the various model levels discussed in the previous chapter.

Second order causes within the economic context

Social

The Polish depression has led to unemployment in the maritime sector. This has a number of social impacts. Firstly, the rise in unemployment will increase the demand for maritime jobs which will enable companies to lower wage rates. Secondly, falling wages combined with rising unemployment represent a decline in affluence and a consequent fall in demand for products and hence shipping. Thirdly, the general economic depression may also lead to social unrest.

Political

The economic climate will place limitations upon political possibilities. One result will be that it will force politicians to place constraints on the maritime sector beyond those which are politically desirable, for example in terms of subsidy reduction, employment levels and development and investment. With elections in mind, such as those of September 1993, the interrelationship of the economic and political elements will become increasingly apparent.

Legal

Economic policy is related to the legal context - for example economic progress requires laws which control bankruptcy, and the level of subsidy allowable. Additionally, economic reform requires that companies must now work within international laws which for example set ship standards. It costs money to maintain the necessary standards to ensure that operations are within the law; however, as they are legal requirements it is imperative that they are obeyed. Meanwhile rules concerning currency convertibility have changed as economic reform progresses and the need to earn hard currency has been reduced considerably.

Organisational

The economic factor of privatisation has numerous organisational impacts. For example it leads to industrial restructuring, competition and diversification. Similarly the closure of many state owned loss making companies will lead to restructuring and rationalisation within companies. This in turn will change markets and change focus within companies.

Managerial

Internal company structure will be affected by privatisation, new market orientation, unemployment and redundancies, competition, the need to survive, and the reduction of subsidies and hiving off peripheral activities. Similarly, the decreased need to earn hard currency will affect company structure and internal operations.

Environmental

There is an economic cost involved with raising environmental standards which will constrain realistic decisions. This involves considerations such as fuel, port pollution, water pollution and lorry controls (noise, air pollution, safety etc). Monetary grants are often used as an incentive by international organisations such as the World bank, International Monetary Fund, and the European Community to encourage countries such as Poland to raise their environmental standards.

Spatial

The economic movement of markets from East to West has recognisable spatial implications. Economic demands have caused a spatial movement of industrial locations - for example economic changes in market demand have caused POL to move their base port from Gdynia to Hamburg.

Logistical

The new competitive economic environment has lead to the requirement for new logistic contexts such as JIT. This, combined with the need to improve reliability creates further economic costs. Generally, market economies require higher standards to survive and integrated services such as warehousing, logistics and quality control. There are also costs involved with entering new markets, and in altering fuel and maintenance sources, necessitated by a new pattern of services.

Technical

The technical aspects of the economic context include the costs of rebuilding and refurbishment. The lack of capital combined with bank controls and difficulties in borrowing finance place severe limitations upon technical development. In Poland the depression has meant that there can only be limited improvements to infrastructure such as roads, railways and ports.

Second order causes within the social context

Political

The social problem of unemployment has political implications in the maritime sector. Changes in maritime industrial conditions such as job and working condition alterations may affect voting patterns and thus the outcome of elections. These aspects are all particularly relevant in specific maritime geographical areas including those around and in Gdynia, Szczecin, and Gdansk.

Legal

Social unrest constrains formation of new legislation, for example on bankruptcies and working conditions in the maritime sector. There must be a balance between that which is feasible and that which is socially acceptable. Another significant issue is the fact that flagging out is now both legal and occurring, and will have considerable social implications for the maritime workforce.

Organisational

It is important that organisational changes are socially acceptable. For example privatisation will result in closures, rationalisation, changing prices and increasing competition. Polish society also needs to accept free market principles generally in the workplace and that the result may be job losses for some, whilst others will earn higher incomes.

Managerial

Changes to the decision making process will give companies greater independence and individuals greater power and control, although they will be constrained by their social acceptability. Managers will have increasing and differing control over workers and all employees will have new roles within the company. This will present many employees with new opportunities. Unions and workers' councils' roles are likely to be depleted and have reduced political impact.

Environmental

The increasing importance given to the green issue means that socially the shipping and ports sector must be seen to be environmentally sensitive. However, environmental improvements will be limited according to society's willingness and ability to pay for a clean environment in competition with financial demands elsewhere.

Spatial

Socially, the workforce will have new mobility requirements and impositions which involve spatial implications. Poland's changing markets will involve new origins and destinations for newly demanded goods. This in turn has implications for port usage and for transport links including shipping, rail and road.

Logistical

Changes to logistics will only be possible if society can cope with, and agrees to, new working practices. The amount of progress made will relate to the willingness of the workforce to accommodate new concepts such as JIT. Another social element with a logistical impact is the practice of hiring out seafarers to foreign shipping companies.

Technical

Poland needs to develop technically to improve its chances of competing in a truly free market. Socially the required construction work may be unacceptable on the grounds of its location or its conflict with environmental interests. Society may also find it unacceptable that Polish ships are to be built, for example, in the Far

East. They may increase the pressure for new buildings to be completed in Poland which may have implications for current and proposed joint ventures. It is also questionable that the use of foreign vessels will be acceptable to the Polish people, and to the Polish maritime workforce in particular.

Economic

Economic policies are constrained by what is socially acceptable, for instance regarding unemployment, and the distribution and level of incomes. Society may also demand the continued subsidy of certain companies in order to help them survive, and successful private companies may be resented to some extent. Foreign companies entering Poland for example through joint ventures may be seen as 'invading' Poland. There is also the complex issue of the relationship between the World Bank/ EC/ International Monetary Fund (IMF) demands, Poland's needs and society's expectations within the maritime sector.

Second order causes within the political context

Legal

Certain aspects of new legislation may be politically unacceptable, for example bankruptcy. State industries may need to be given some legal protection to satisfy the electorate. There is added political pressure to pass legislation to satisfy the requirements of international organisations such as the World Bank/EC and the IMF in order to gain much needed loans and grants.

Organisational

Privatisation requires organisational changes within a company which may have indirect political consequences. For example rationalisation leading to job losses and income reduction make privatisation a difficult political choice. Hence political issues may constrain organisational decisions.

Managerial

Changes in the managerial element will have small scale political connotations and may create local voting issues. There may be political power struggles within companies as management and worker structures change thus affecting, and potentially constraining, managerial change.

Environmental

The environment may be used within the maritime sector as a political issue which influences voting. Politically it may be seen as important to promote environmental awareness and to control emissions, fuels and safety. However, reaching and maintaining new standards may prove costly and could lead to job losses. The political and environmental pressures need to be carefully balanced at every level. Internationally the requirements of the World Bank/ EC/ and IMF will create extra pressures.

Spatial

Political persuasions and votes usually follow a spatial pattern. In Poland the maritime sector is a major employer in restricted areas and as such is of political significance spatially. Changing markets will change spatial impacts in terms of port usage and rising or falling production levels in different regions.

Logistical

New logistical practices will have political impacts depending upon their acceptability, for example, new working practices required by new logistical standards may prove unworkable. They may mean votes being lost or won so that logistical changes may be influenced to some extent by political considerations.

Technical

Partial concurrences between the technical and political contexts depend upon whether technical changes - such as upgrading road and rail links, building ships abroad, revamping or closing ports and shipyards and reducing the workforces - are politically acceptable.

Economic

There is a clear economic impact within the political context. Politics dictates what is acceptable in the Polish Parliament, to the electorate, to workers and management, and internationally with regards to economic issues including closures, subsidy, investment pricing, privatisation and bankruptcy. Politics and voting patterns dictate economic progress.

Social

Changes in the political context have an impact upon society. Recently these have included unemployment and the level and distribution of income which have altered partly through the introduction of privatisation and joint ventures. Partial convertibility of the zloty and dramatic exchange rate fluctuations have also had social consequences, eroding savings and making the acquisition of hard currency unnecessary.

Second order causes within the legal context

Organisational

Changes in the legal context will have organisational impacts. Such legislation will include the creation of competition, foreign investment, bankruptcy, joint venture and privatisation laws, as well as laws concerning state involvement and subsidy. There will be adjustments to the current banking and international trade law and the introduction of laws which govern aspects concerning hard currency such as exporting, importing and rates of exchange.

Managerial

Legal aspects which will have impacts upon the managerial context include changes to internal company law, for example governing workers councils and the involvement of workers in individual decision making. There will also be new private company law controlling items such as accounts and reports. The appointment and responsibilities of directors will also change considerably.

Environmental

The legal system provides the basis for environmental control. New legislation has already been developed but this needs to be expanded and developed further. The new laws set new environmental standards which must be adhered to. These will concern the setting up of monitoring systems. Progress on the environmental issue can only occur as quickly as legislative progress is made through parliament.

Spatial

Laws define spatial maritime aspects such as port boundaries and territorial waters. Other aspects which might be taken into account here are the boundaries of various constituencies and the potential effects of any changes to them.

Logistical

The legal context defines allowable levels and types of relationships between companies through, for example, monopoly legislation, thus encouraging or inhibiting vertical or horizontal integration which is a major feature of modern logistical practice moving into Poland.

Technical

Technical aspects must also come into line with legal requirements. These aspects include ship condition, port activities, safety rules - especially as regards manning, ports, road and rail vehicles. Legal constraints require more technically than is feasible in the short term. International law has legal connotations - for example regarding vessels. Pressure upon maritime operators to raise technical standards needs to be backed by law to ensure that action is taken.

Economic

Laws control the pace of economic reform. Economic progress takes place in line with the introduction of for example bankruptcy laws, banking laws, subsidy laws, and privatisation laws. Other legal aspects with an economic impact are the controls over wages and laws upon joint ventures and foreign investment.

Social

Laws directly affect social issues by imposing conditions relating to unemployment, wages, and standards of living. Similarly, the freeing of markets, prevention of subsidy and the enabling of bankruptcy have direct and obvious social impacts.

Political

The legal framework affects politically sensitive issues, such as unemployment and wages. Laws must reflect political ideals and take into account the reaction of the electorate.

Second order causes within the organisational context

Managerial

Privatisation in the organisational context influences management structures by altering responsibilities and targets for achievement. It also affects the numbers and roles of workers, types of work, as well as unions and worker/government relationships. Organisational changes have enabled management to diversify into other industries. Changes in the marketing environment have created the need for new management styles and skills which can be achieved through training.

Environmental

New organisation of the market system affects environmental considerations and highlights the contradictions of profit motives and environmental protection. Removal of state control from many industries will have environmental implications. Private companies are in business to make profits, and are usually less concerned with public environmental issues than the state. Also the introduction of a free market will create a multitude of small companies with a lack of coordination on broad issues such as pollution.

Spatial

Changes in organisation structures and markets will change the spatial distribution of activities. This may involve a change in the choice of mode, infrastructure and ports. Further, the introduction of joint ventures with foreign interests and collaboration with other industries, such as warehousing, will have spatial impacts.

Logistical

Organisational changes will allow new logistical patterns to emerge as new objectives are faced such as JIT and integrated services. There will be no state direction of traffic as the coordinating function of the state is removed and industry acts independently.

Technical

New organisations will be required to maintain technical improvements to achieve success in the maritime market place. The technical improvements necessary are likely to include better ships and equipment to foster greater reliability and faster turnaround times. The potential change in the reliance upon certain modes would also require considerable technical alteration. Joint ventures and foreign investment may facilitate change though the provision of capital, unavailable domestically.

Economic

The organisational change of privatisation has major economic impacts. Foreign and Polish shareholders needs must be taken into account. Sources of investment both foreign and domestic are changing rapidly. Other economic impacts of organisational change will include changes in income, the introduction of competitors, and bankruptcy. Further concerns include the formation of joint ventures and the ensuing drain/input of capital and profits and the flow and direction of hard currency earnings.

Social

Privatisation involves a number of social impacts. These include job losses, creation of new but different jobs, changed incomes and new responsibilities. Joint ventures will introduce similar social impacts but in addition will create social concerns over the invasion of foreign companies. Further organisational change with social overlaps include the effects on seafarers of using flags of convenience. There will likely be changes in attitudes towards part time workers, and women, and a possible loss of social facilities.

Political

The organisational issue of privatisation will have a political impact in terms of its acceptability to the work force. Joint ventures and foreign investment will have a political impact as will the degree of acceptance of state withdrawal from industry and the consequential threat of bankruptcy and drive towards efficiency.

Legal

Organisational changes such as joint ventures, hard currency earnings and foreign investment require enabling legislation. International bodies such as EC/ World Bank/ IMF create further pressure for legislation to be passed to encourage privatisation. This will require laws on take over, director responsibilities, ownership, dividends and numerous other issues.

Second order causes within the managerial context

Environmental

New managerial needs and demands may lead to environmental neglect, especially as privatisation will focus management on profit earnings. Further problems emerge as staff numbers are reduced conflicting with the introduction of the environment as a new issue perhaps requiring more staff with new specialities.

Spatial

Managerial change will have a spatial impact in terms of reductions and alterations in services, routes and offices, with some of the latter moving abroad. Consolidation of locations is also likely.

Logistical

Managerial and logistical contexts integrate with respect to the ability of new management to cope with the logistics concept, and their level of understanding of JIT needs. More specifically, they will need to have the ability to negotiate new contracts and to cope with joint ventures, foreign investment, and competitors. The ensuing new logistical requirements may necessitate management training.

Technical

Management needs to recognise new technical apparatus and techniques and their necessity. They must be able to cope with technical change emerging from such areas as information technology development. This will require some extensive training and education.

Economic

Management must have the ability to conduct the economic side of business within the new climate. They must cope with new accounts and negotiate contracts. New management skills made necessary because of economic change include decision making, understanding of profits, markets, niches and competitors' economics. There will be new economic opportunities, especially through the creation of joint ventures and foreign investment, as well as a need to understand foreign partners' economics and deal with changes regarding the hard currency issue.

Social

Socially, workers will need to cope with new roles, new management and new workmates. Control and responsibilities in the workplace will alter and changing incomes will have further social implications.

Political

Managerial change resulting in new roles for the workforce will have political implications in terms of voting. The political power of workers within companies generally is reducing due to the falling membership and fading control of unions. This results in a corresponding loss of influence over local and national government decision making and policy.

Legal

Laws are influenced by previous management arrangements and need to change in the context of the new environment. Eventually there will be new laws relating to worker power, party control, state influence, bankruptcy and management responsibilities.

Organisational

The potential of organisational changes to be effective, for example privatisation, joint ventures and foreign investment, are constrained by workers, managers, directors and the company's internal hierarchy.

Second order causes within the environmental context

Spatial

Companies such as POL may be under pressure to change feeder modes - for example from truck to rail - to improve their image environmentally. There may be environmental pressure to change port locations or to change fuel sources, and environmental awareness may lead to demands for new products or new product sources. Each of these changes will have a spatial impact. Alternatively, demands for greater environmental awareness through spatial change may come from abroad, or for example from joint venture partners or foreign investors.

Logistical

New routes, chosen for environmental reasons, direct or indirect, will affect logistics. Environmentally there may be pressure to change warehouses used, to change modes or to improve current modes of transport such as lorries. Each of these aspects will have a logistical impact.

Technical

Environmental pressures are likely to have considerable technical impact, for example on ports, ships, and related transport modes in terms of operations, fuel, and maintenance. Pressures to raise environmental standards will come from foreign investors, joint venture partnerships, EC, World Bank, IMF, and other international bodies, and from within Poland. However, the amount of progress that can be made in the short term is limited considering Poland's starting point.

Economic

Higher environmental standards generally result in higher prices which may be reflected in fuels, ship condition, maintenance, port and other operational issues.

Pressures for improvement come from numerous sources including foreign bodies such as the World Bank/ IMF/ EC who will only supply grants and loans if green standards are being met. However, these pressures have to be balanced with the limited amount of finance available.

Social

Environmental issues change job requirements within the maritime sphere, but the increased costs involved with raising standards may lead to cost cutting exercises which include job losses. New standards of workmanship will require training, new ideas and new responsibilities. Local environmental improvements will provide societal benefits, but will raise the price of goods.

Political

A poor environment has political repercussions. Nationally it may mean loss of popularity with the electorate, and internationally it may damage the country's reputation and possibly delay entry to the EC even further.

Legal

New environmental standards require new national laws. Compliance with EC laws places new demands on shipping. Port and other infrastructure is directly affected by the environmental context. Laws restricting other industries' choice of resources to green products will affect imports and exports.

Organisational

Where these two contexts integrate it must be considered whether environmental needs can be incorporated into new, free market industries without the central control and encouragement of government.

Managerial

New environmental issues require new management techniques, skills and training and will involve new jobs with new roles.

Second order causes within the spatial context

Logistical

Logistical patterns are partly determined by the location of sites of production and consumption, and by infrastructure, ports and markets. Ship types and the constraints upon their usage and the availability of equipment in differing locations may determine the potential of logistical improvements. Other areas include the new emphasis on western markets, EC border controls, location of competitors and warehouse availability.

Technical

The location of ports and infrastructure within the spatial context may affect the ability to improve maritime and transport technology. The distribution of existing technology and the availability of land for its improvement provide further examples of technical and spatial overlaps. Also, the origins of joint ventures and foreign investments will affect the technical standards required and the availability and extent of investment.

Economic

Economic development and change will occur according to where investment is placed. Spatial factors with an economic overlap include the location of new road and rail links, port location and development. Spatial changes in the market will result in economic benefits and costs in new areas. There may be spatial changes in the location of industry as sources of consumption and production are moved to areas of greater economic advantage, for example less expensive sources of fuel.

Social

Spatial and social factors occur where the creation of new opportunities leads to a change in the distribution of employment. Incomes will vary according to where investment takes place which also has impacts upon consumption and production. Socially, companies such as POL need to retain presence in certain areas, such as Gdynia, for political reasons. Other issues for consideration include flagging out, the spatial impact of seafarers from elsewhere and in particular overseas, movement of ships to other ports and changes to routes.

Political

Politically the decline of the maritime industry in certain areas such as Gdynia, may not be acceptable and hence there may be spatial constraints on political developments. Other spatial changes such as changing modes, closure of ports or production facilities will be affected by their level of acceptability, politically. Foreign investment and joint ventures will also have spatial impacts with political implications, as will moves towards the EC and changes in the routes served.

Legal

The spatial movement of shipping into foreign markets will affect the legal requirements of ships in terms of quality, condition and manning. The use of foreign ports has further legal implications. Poland's moves away from East Europe generally and towards becoming a member of the EC will have a significant legal impact at every level of maritime operation.

Organisational

Organisational changes such as privatisation, joint ventures, and foreign company involvement may lead to closing down of branches, with some movement of responsibilities abroad. This may result in new routes and changes in ports of call as companies aim to serve new markets and find new sources of goods.

Managerial

Spatial and managerial aspects occur where there are new job locations and the work force generally needs to be more mobile and flexible. Other changes have been created by the use of flags of convenience, new routes and other infrastructure and by varying crew origins. New managerial skills are required to cope with the spatial changes.

Environmental

Environmental pressures vary between areas. The spatial changes in terms of new ports, and new routes may create new environmental stresses and problems. The use of rail as opposed to roads invokes numerous environmental arguments and redistribution of impacts.

Second order causes within the logistical context

Technical

New logistical practices demand new facilities such as specialised warehousing, as well as improvements to existing technical facilities such as ports, road, rail, lorries and ships. With improvement these facilities will enable greater reliability and better levels of service to be achieved. IT and communications play a vital role here.

Economic

The adoption of new logistic practices should have economic effects in terms of improved incomes, reduced waste, reduced costs and improved cash flow. General industrial and economic improvement could lead to greater demand. However, logistical improvements also lead to job losses causing higher unemployment and falling incomes which in turn lead to lower demand. Similarly, the transfer of economic investment abroad perhaps through joint ventures may lead to loss of activity within Poland.

Social

Increasing use of modern day logistical techniques can lead to reduced manpower, less income and consequential social problems. They also call for changed working practices, roles, responsibilities and new skills, all of which are social impacts. However, some workers may benefit from extra income and there may be price reductions and more flexibility in the markets.

Political

A number of the logistical changes which take place will have political implications, for example changing sources of production and consumption and the use of different modes of transport. The falling availability of jobs, combined with their changing roles and locations also have political and logistical impacts. Further, foreign influence over logistical changes may not be politically acceptable.

Legal

There is a requirement for the introduction of new laws to allow the market to become free enough for new logistical practices to develop fully. These include laws allowing foreign investments and joint ventures as well as facilitating easier currency movements and governing repatriation of profits.

Organisational

Logistic development places demands upon various industries to coordinate with one another in an organisational context. This raises questions as to whether newly privatised companies, joint ventures and foreign companies working in Poland will coordinate fully within new market conditions.

Managerial

Management needs to be able to cope with new logistical operations and this may require training to develop the necessary skills. New internal structures and a flow of information are required, as is an ability to cooperate with personnel in other companies.

Environmental

Logistic changes such as the likely move toward road transport have significant environmental implications. There are several environmental arguments concerning the use of rail versus road, and the use of ship versus road. Other problems are likely to emerge where pressure upon finance encourages short cuts which are detrimental to the environment.

Spatial

Logistical and spatial contexts occur where there are changes to the infrastructure, storage, warehousing, sites of production and consumption, as well as routes and ports used. These logistical changes will have significant spatial effects.

Second order causes within the technical context

Economic

It is likely that existing technology will fail to allow the degree of economic change which is desirable. Further, technological needs, such as new ships and roads, place a drain upon the economy of such scale that they are unlikely to be achieved within the foreseeable future.

Social

Technological development of ships, ports, roads and so forth should improve working conditions. It will also require changing roles and new skills in the workplace. Other technical and social mergers occur through the reduction in incomes and jobs leading to rationalisation and through a change in the distribution of income.

Political

The level of investment necessary to raise technological standards sufficiently may not be available if the electorate is not favourable to the idea of extra public expenditure and hence higher taxation. Similarly, the development of new roads and the closure of some port facilities will have local political impacts. Other factors include the technological influence from joint ventures and foreign companies, such as the introduction of new developments and competition, and their political acceptance.

Legal

The introduction of tighter laws governing technological standards on matters such as ship maintenance, fuel, roads and vehicles may prove pointless if technology is unable to keep up. Additionally, technological developments place a demand upon the legal process at local, national and international levels.

Organisational

Organisational changes such as privatisation may not be feasible if the technical side of the industry is inadequate. This is due to the fact that investors are unlikely willingly to become involved in an industry which is technically lacking, and consequently, without investment, the company will not survive.

Managerial

Management needs to be capable of coping with new technology, which is likely to require new skills and training, and changes to personnel. The attitude of employees is important.

Environmental

The introduction of new technology should reduce environmental damage. However, if capital is not available for replacement, the continued use of old equipment, now unsubsidised, will compound the problems. Further, technical change may mean a switch of modes to one which is more environmentally damaging, such as road.

Spatial

Technical changes such as use of new ports both foreign and domestic, changed use of infrastructure (eg rail to road) and changed production and consumption sites will all have spatial impacts. New equipment may enable new trade routes to be opened, whilst the technological development of communications and IT will have further spatial implications.

Logistical

Technical improvements will affect the potential of logistical developments. For example, this may be achieved by introducing new warehousing and enabling activities to be better integrated. IT and communications development will make possible new methods of logistics such as JIT.

The Contextual Matrix Model is shown in figure 3. Clearly some cells contain numerous overlaps whilst others remain relatively empty. It is suggested that those issues which are most important to Polish shipping are usually those that occur most often within the model - ie. in the highest numbers of boxes. Clearly this is not always the case, as some issues may occur only rarely and yet be fundamental to the development of Polish shipping - and vice versa. As a result, some qualitative interpretation had to be used to ensure that other major issues are not overlooked.

   Whilst recognising that there remain difficulties with prediction in the shipping sector, in Poland and worldwide, by using a model based upon a pragmatic philosophy it is suggested that the main changes in the political, social and economic areas might be identified in a contextual parameter and from the matrix model as the following:

Political context
>   Democracy/ decision making
>   Joint-ventures/ western involvement and investment
>   Competition policy/ state withdrawal
>   Privatisation/ state withdrawal

Economic context
>   Hard currency
>   Joint-ventures/ western involvement and investment
>   Removal of subsidies
>   Privatisation/ state withdrawal

Technical context
>   State of Poland's fleets
>   State of infrastructure
>   Communications network

Managerial context
>   Decreasing government involvement
>   Labour - availability and level of skills

Social context
>   Unemployment
>   Free market pricing

Contexts	Economic	Social	Political	Legal	Organisational
Economic		Unemployment leads to increased demand for jobs, lower wages and fall in demand for products. Economic depression causes social unrest.	Economies may effect political decisions and political processes, for example attempts at raising employment levels.	Laws control bankruptcy, subsidy, ship standards etc. - which create costs. Hard currency issues. Condition of economy facilitates local change.	Privatisation driven by economic need leads to restructuring, competition, diversification, closures and rationalisation, changing markets.
Social	Acceptability of society constrains income, unemployment and subsidy removal. Foreign companies 'invading'. Relationship of WB/IMF/EC demands, Poland's needs and society's expectations.		Unemployment and changes in social conditions - especially in certain areas affects voting patterns.	Society's level of acceptability constrains new laws eg. working conditions, bankruptcy, working conditions, flagging out.	Social acceptability of changes such as privatisation meaning closures, rationalisation, price changes and competition, free market principles, job losses, changing incomes.
Political	Politics dictates what is acceptable with respect to closures, subsidy, investment, pricing, privatisation, bankruptcy.	Unemployment, level and distribution of income, changes through joint ventures and privatisation. Zloty convertibility and exchange rate.		Acceptability of new laws eg on bankruptcy, pressure from WB, IMF EC to pass laws to gain grants.	Privatisation leads to rationalisation, job losses and falling income - political acceptability?
Legal	Law control pace of economic reform eg bankruptcy, banking, subsidy, joint ventures, privatisation, wages controls foreign investments.	Laws impose conditions upon unemployment, wages, standards of living, freeing of markets, subsidy, bankruptcy.	Laws on sensitive issues eg unemployment / wages, must reflect political ideals and consider electorate reaction		Competition, foreign investment, bankruptcy, joint ventures, privatisation, subsidy, state involvement, banking trade, hard currency aspects all have organisational impacts.
Organisational	Privatisation means share holders needs are important. Changes in sources of investment, income, competitors, bankruptcy, joint ventures, drain/input of profits, flow of hard currency earnings.	Privatisation involves job loss and creation, changed roles and incomes. Joint ventures create invasion worries. Other issues include flags of convenience, part time workers, women, social facilities loss.	Acceptability of privatisation, joint ventures, foreign investment, state withdrawal, and threat of bankruptcy.	Changes such as joint ventures, hard currency and foreign investment require enabling legislation, further pressure from EC/WB/IMF to encourage privatisation requires numerous laws.	

Figure 3   The contextual matrix model (top left section)

Contexts	Economic	Social	Political	Legal	Organisational
Managerial	Management ability to cope with accounts, contracts, profits, decisions, markets, competitor economics, and new opportunities through joint ventures/ foreign investment. Hard currency changes require training.	Workers need to cope with new roles, new workmates and changes in control.	New roles in the workplace may affect voting. Political power of worker less due to fading unions, loss of worker influence over government.	Changes require new laws relating to worker power, party control, state influence, bankruptcy, management responsibilities.	Potential of privatisation joint ventures, foreign investment, constrained by employees and companies' internal hierarchy.
Environmental	Higher standards mean higher prices, eg fuels, ship conditions, ports, maintenance. Pressures from EC/ IMF /WB, who will only supply grants of standards are met. Limited finance to achieve standards.	Job requirements change, new standards require training, cost of raising standards may lead to cost cutting and job losses. Local improvement benefits society but raise prices	Poor environment may mean loss of popularity with electorate, and may damage international reputation and delay EC entry.	New standards need new laws. EC laws will place new demands. Laws restricting industrial choice of resources affect imports and exports.	Can environmental needs be incorporated into free market industry without control of central government ? Can industry be structured to accommodate it ?
Spatial	Location of investment, new road /rail links, port location/development, changes in markets and location of industry in areas of greater economic advantage.	New opportunities change employment distribution; incomes vary with investment location, affecting consumption; companies need to retain presence in certain areas; flagging out, foreign seafarers, changed routes.	Level of acceptability of decline of industry in certain areas, change of modes, port closure, foreign investment, joint ventures, moves towards EC, changed routes served.	Moves into new markets and ports affects ship requirements spatially. Move from East Europe to EC has legal implications at every level.	Privatisation, joint ventures, foreign involvement, may lead to closures, companies moving abroad, new routes/ ports/ markets served.
Logistical	New practices should improve incomes, and cash flow, reduce waste and cost, may lead to greater demand, job losses, lower incomes and lower demand, transfer economic activity abroad, lead to increased costs due to raising company standards.	Logistic techniques can lead to reduced manpower, less income, call for changed working practices, responsibilities and skills; some workers may gain extra income and there may be price reductions and flexibility in markets.	Changing sources of production/ consumption. Use of different modes, falling job availability, changing roles and locations, level of acceptability of foreign influence over changes.	Need new law for logistics to develop eg allowing foreign investments and joint ventures, easing currency movements, and controlling repatriation of profits.	Development demands coordination - will new ventures coordinate fully in new markets ?
Technical	Technology may fail to allow economic change; needs such as ships and road building drain the economy.	Technical development improves working conditions, requires new roles, skills and training. Income and jobs rationalisation.	Electorate's acceptability of investment in technology, new roads, port closures, influence of foreign companies.	Technology must keep up with new laws eg. on maintenance, fuel, ships. Developments place demand on the legal process.	Technical change requires new organisational structures and communications - IT.

The contextual matrix model (bottom left section)

101

Contexts	Managerial	Environmental	Spatial	Logistical	Technical
Economic	Company structure changes due to privatisation, market changes, unemployment, competition, loss of subsidy.	Economic cost of raising standards eg fuel, pollution, lorry controls, grants used as incentive by WB/ IMF/ EC.	Move of markets from East to west. Movement of industry, POL moved base port from Poland to Hamburg.	Costs of entering new markets and changing resource supplies. Economics demands new logistics in order to be competitive.	Costs of rebuilding and refurbishment. lack of capital, bank controls, borrowing difficulties limit development.
Social	Greater independence of companies and individuals - new roles, changing control and opportunities, unions less important.	Maritime sector must be seen to be sensitive - limited improvement due to willingness to pay.	New mobility requirements of workforce. Changing markets mean new origins and destination, affecting port usage and transport links.	Capability and willingness of workforce to accept new practices. Hiring out seafarers to foreign companies.	Acceptability of improvement and building vessels abroad etc. Use of foreign vessels.
Political	Local voting issues. Political power struggles within companies.	Issues influence votes. Politically important to promote awareness and control pollution and safety. Acceptability of costs involved ? Pressures from EC/ WB/ IMF.	Maritime sector of political significance spatially. Changing markets created by political change affect port usage and rising/ falling production.	Acceptability of new logistical working practices.	Acceptability of changes eg upgrading road and rail links, building ships abroad, closing ports, job losses.
Legal	Changes to internal company law eg workers councils. New private company law eg reports and accounts, director appointment.	Law is basis for control and dictates progress. New laws will include setting up monitoring systems.	Laws define spatial aspects eg port boundaries and territorial waters.	Law defines allowable relationships between companies. Polish law defines degree of rail involvement in shipping.	Ship condition, port activities, safety - law dictates technical standards - are these feasible ?
Organisational	Privatisation alters responsibilities and targets, number and role of workers, unions and government relationships, diversification possible; market requires new skills.	Profit versus the environment. Removal of state control and introduction of private companies, lack of coordination of numerous small companies.	Changes in spatial distribution of activities eg mode, ports, infrastructure, joint ventures, foreign interests collaboration with other interests.	Organisational change allows new logistics eg JIT. No state direction of traffic, industry sets independently.	New organisation necessitates new technology eg better ships, and reliability, faster turn around. Potential change of modes, joint ventures, foreign investment.

The contextual matrix model (top right section)

Contexts	Managerial	Environmental	Spatial	Logistical	Technical
Managerial		New managerial needs and focus on profit efks privatisation may mean environmental neglect. Environmental aspect requires more staff, organisational context less.	Reductions and changes to services, routes and offices, eg POL base now Hamburg, consolidation of locations.	Ability of management to cope and understand new contracts, joints ventures, JIT, foreign investment, competitors, logistics - may require training.	Management must recognise and cope with new technology - may need training.
Environmental	Environmental pressures require new management techniques, skills and training for new jobs.		Environmental pressure to change modes, port locations, fuel sources, demands for new products or product sources. Pressure from abroad via joint venture investors etc.	Environmental pressure for new routes, modes, warehouses, or improvements to current modes etc.	Environmental pressures on ports/ ships etc, regarding fuel, operations, maintenance. Pressure from foreign investors, joint ventures, IMF/ WB/EC and within Poland. Progress limited by 1992.
Spatial	New job locations needs more mobile workforce. Flags of convenience, new routes/ infrastructure, varying crew origins, all require new skills.	New ports/ routes etc may create new environmental issues. Rail versus road.		Logistics determined by production/ consumption, location, infrastructure, ports. Ship types and availability determine logistics potential. Emphasis on west markets, EC border controls, location of competitors.	Port/ infrastructure location affects ability to improve - distribution of technology and availability of land. Origins of joint ventures and investors influence standards required.
Logistical	Management's ability to cope with logistics - new internal structure, information flows, cooperation with other companies and training.	Likely move to road transport. Ship versus road, rail versus road. Pressure on finance could lead to damaging short cuts.	Changes to infrastructure, storage, warehousing, production/ consumption, routes and ports served.		New practices require new and improved facilities, eg ports, road, rail, warehouses to provide better service levels, IT and communications.
Technical	Ability to cope with new technology - new skills and training, personnel and attitude changes.	New technology should reduce damage, but continued use of old, unsubsidised equipment will compound problems.	Use of new infrastructure and changes in production/ consumption sites. Development of IT/technology may open new routes.	Technical improvement dictates logistic potential eg integrating activities, new warehousing, IT and communications development.	

The contextual matrix model (bottom right section)

Legal context
Polish law changes eg. bankruptcy
Application of international law

Organisational context
Privatisation/ state withdrawal

Logistical context
Changing production techniques
Increasing western influence

Environmental Context
Increasing pressure from foreign interests and international bodies

From this summary several main issues stand out which will have a serious and significant impact on commercial and operational activity in the Polish shipping sector and which are both complex and interrelated. These may be summarised as :

- Privatisation / state withdrawal
- Joint ventures/ western industrial intervention and cooperation
- Technical condition - ie. fleets and their renewal, infrastructure and communications
- Democracy and labour issues

We now turn to greater detail of these issues in the context of the maritime sector of Poland.

Privatisation

Privatisation or denationalisation is the removal of at least the majority part of state control or ownership from industry. Although far from exhaustive or all embracing this definition is adequate for our purposes.

Reasons for adopting a privatisation policy in any economy are often varied but can include the following (Foreign Trade Research Institute 1992):

1   It promotes economic freedom.
2   It enhances efficiency.
3   It eases the problems of managing public sector employees and in particular pay demands.
4   It reduces the need for public sector borrowing.
5   It is often seen as a politically attractive policy with the electorate.

For East European countries, privatisation is seen as the linchpin of transition to a market economy (International Management 1991), and although it is progressing slowly, it looks likely to continue. The motivation for privatisation in East Europe is slightly at variance with the reasons given above and includes the rather more specific need to raise cash to pay pressing foreign and internal debts, and the belief that privately owned companies will be more willing to invest in the new technology needed to raise efficiency and enhance competitiveness (Lloyds Shipping Economist 1990).

According to Lloyds Shipping Economist, Oct 1990, problems with privatisation include:

- lack of viable funds (ie. transferable currency) within East European countries to buy state companies. This problem is exacerbated by the extensive international debt of the region. For example, in 1990 Poland's debt stood at $43.4bn, Hungary's at $20bn, ex-Czechoslovakia's at $6.8bn, the then Soviet Union's debt being $54bn;
- difficulties in valuing companies where there is no history or tradition of meaningful valuation of assets, although new accounting procedures should help;
- suspicion of foreign capital;
- level of western interest remains to be seen; the slow rate of privatisation up to 1990 may have been partly due to feelings of political or economic distrust;

Despite these problems East European countries appear to be continuing actively with privatisation. Examples of progress can be found from each of the East European states, but we will look specifically at the Polish situation.

Poland

As this research uses Poland as a case study, Polish privatisation progress will be considered in some detail. Although a draft privatisation law was adopted by the Council of Ministers on 5 March 1990, enabling legislation only came into force in July 1991 (ESCEC 1990, Lloyds List 10/9/91).

According to the Polish Act on the Privatisation of State Owned Enterprises, passed on 1st August 1990, the privatisation of such an enterprise is based on offering to third parties, shares or stocks of a company evolving from the transformation of a state owned enterprise and owned exclusively by the state treasury, or offering to third parties the assets of a state owned enterprise or the sale of the enterprise (Centre for Privatisation 1990). According to Article 23 of the Act, shares belonging to the state treasury could be sold in three main ways:

1   on an auction basis

2   on a public offer basis
3   on a negotiation basis after a public invitation

Article 24 adds that employees of a state owned enterprise transformed into a company are entitled to buy up to 20% of the total amount of shares of the company held by the state treasury on a preferential basis. Also shares sold on a preferential basis to employees should be sold at 50% discount compared to the price set for 'natural persons'. Article 25 stated that Parliament, on a motion of the Council of Ministers, 'shall pass a resolution regarding the issue and value of privatisation coupons which can be used to pay for:

1   acquiring shares issued as a result of the transformation of a state owned enterprise
2   acquiring title to participation in financial institutions which will have at their disposal shares created as a result of the transformation of state owned enterprises
3   acquiring enterprises or integrated parts of the assets of state owned enterprises referred to in other articles'

The privatisation coupons were to be distributed free of charge, in equal amounts to all citizens resident in the country. The rules concerning the usage of such coupons were to be determined by the Council of Ministers.

Also passed in Poland on 1st August 1990, was the Office of the Minister of Ownership Changes Act, which outlined the duties of the Minister of Ownership Changes. According to the Act the Minister should implement the state's policy on ownership changes and in particular should:

1   'prepare guidelines on state policy on the privatisation of state owned enterprise
2   prepare guidelines on state policy on capital co-operation with foreign partners
3   carry out tasks specified in regulations on the privatisation of state owned enterprises
4   make analyses of the state of ownership changes
5   co-operate with trade unions, associations, chambers of commerce and other civic organisations and with bodies of state administration and local government in the field of formation and development of private enterprises
6   initiate personnel training and further education in the fields of privatisation activity, capital markets and development of private enterprises, and publicise experiences and information in these fields' (Centre for Privatisation 1990).

Under the terms of the Balcerowicz programme (named after the former finance minister) some 400 state owned enterprises were to be sold in the first wave of privatisation. Finding buyers for such a sizeable sell off was likely to prove difficult, although the government appeared keener to dispense with assets at almost any price rather than be forced to support loss making industries. Balcerowicz favoured management buy out as the means of transferring state firms into the private sector. This, he believed, would concentrate ownership and would lessen the dangers inherent in all-out employee share ownership. Opting for management buy outs was also preferable since there seemed to be few foreign investors, a situation which was disappointing after all the encouragement from the West to leave Communism. However, by October 1991 80% of the retail industry was in private hands, with the total number of private businesses standing at around 1.1 million. Three thousand companies had been set up with foreign capital worth about $400m and it was hoped to privatise 40% of the economy within five years (Wall Street Journal 28/10/91). However, by the end of April 1992 privatisation revenues had reached only Zl 600 bn, raising questions as to whether the Zl 10.000 bn year end sales target was attainable. The privatisation ministry planned to put around eight companies on sale through public share offers linked to sales of large share holdings to foreign investors (Finance, East Europe, 1992), 50 - 60 enterprises were to be privatised through trade sales and privatisation would extend to between 400 and 600 enterprises.

The most common method of privatisation to date has been through liquidation and the first taste of market forces for many state owned enterprises has proved their last. 'Classic' privatisation on the UK model and privatisation by voucher have also been used. To these should be added the rent of public assets or so called invisible privatisation where concerns are simply taken over by those running them.

The government's privatisation programme outlining main directions of privatisation undertakings in 1992 also included the following (Economic Review 22/05/92):

Commercialisation The commercialisation or 'corporisation' of state owned companies was to continue with around 250 enterprises singled out for the process.

Reprivatisation Reprivatisation is the return of state requisitioned assets to their original private ownership. The government declared that it considered the settlement of this issue the most urgent task of the privatisation programme. This included creating a legal basis for reprivatisation and separating assets with which to settle reprivatisation claims.

Mass privatisation programme Scheduled to commence in 1992 this would continue throughout 1993 when investment share certificates would be distributed among citizens. The programme was seen as having a number of merits. It would further the restructuring of the economy, offer capital access to a sizeable number of enterprises at the same time, and would have the support of privatised enterprises workforces', as 10% of new companies' shares would be distributed among the workforce without payment.

Project sectoral approach

Under one of the foremost projects promising to improve the efficiency of Poland's financial sector, nine state owned banks were to be privatised. Plans were also made to privatise the foreign trade sector.

Jacek Siwicki, then Secretary of State at the Ministry of Privatisation, favoured the sectoral approach to privatisation, having decided that the advantages of this approach far outweighed the disadvantages. The sectoral approach leads to various privatisation and restructuring paths. Sectoral projects are divided into three phases. In the first phase, an analysis of a specific branch of industry is undertaken. In phase two, restructuring and privatisation strategies are developed for enterprises within the sector. In the final phase the strategy is implemented. One advantage of this approach is that the privatisation decisions are taken following a thorough analysis of the industry both in terms of its position in the domestic market and its influence on the international market. The Ministry of Privatisation also has more control over the future of the entire sector because this approach enables them to package weaker companies with stronger ones. Another plus is that the industry appoints one advisor to work on the group of enterprises. This has benefits as there is a transfer of experience from each case and unit cost is cheaper. The sectoral approach gives the Ministry the opportunity to assess the impact of individual privatisation transactions on the industry as a whole (Pole Position 1991).

Proceeds of privatisation by sale of equity were estimated at Zl 1.2 trillion. Privatisation by liquidation was expected to generate Zl 2.8 trillion and privatisation of banks Zl 1.2 trillion. However these estimates of revenue deriving from the sale of the Treasury's assets were difficult to achieve and sometimes erroneous.

Maritime sector

In the maritime sector of East Europe as a whole it is thought that shipping could be affected by privatisation at a relatively early stage because it may be one of the easier areas to privatise. Reasons for this include its attractiveness to westerners as it involves large capital assets with markets attached; also shipping is in less financial difficulty than many other industries. However it is sometimes argued that privatisation may not occur in this sector in the immediate future as East Europeans generally do not have enough money to buy into such a capital expensive industry, and where foreign companies were allowed to buy it is doubtful whether they would want to, due to the limited profits being generated.

Despite this, examples of privatisation of the East European shipping industry do exist in Poland, where the Government has included POL, PZM, the ports of Gdansk, Gdynia and Szczecin ship repair yards and fishing companies in the list of 400 companies which will be privatised as a priority (Fairplay 15/8/91). It is believed that privatised shipping industries would be more willing to invest in new technology needed to raise efficiency and enhance competitiveness (Lloyds Shipping Economist 1990).

The Polish shipyard Stocz Gdanska has been partially privatised and 20% of the shares sold to the workforce. The Polish Minister for Industry announced on 10th April 1990 that the yard would be an experiment in privatisation. It had turned into a public limited company, with 80% of the shares held by the government. The 20% of the shares offered to workers was initially rejected causing the government to revalue the share offer at 30% less. By May 1990 about one third of the work force were reported to have bought shares (Fairplay 10/5/90).

Szczecin shipyard was given two years in which to complete a sale, having worked out a privatisation strategy. Its position is much better than many Polish companies as the commercial manager Mr Zarnoch pointed out 'shipbuilding could always earn valuable foreign exchange and it was treated favourably by Warsaw'. Another attraction for investment is the yard's extensive order book, which covered 75% of the shipbuilder's available capacity for the next three to four years and was valued at around $800m in September 1991.

Centromor, the main state held ship export agency, is likely to be the first organisation among Poland's shipping industry to be sold off, having made the first move towards privatisation in September 1991. The floatation method favoured by managing director Mr Ferworn, would be for a foreign owned bank or financial institution to take a controlling stake in Centromor (Lloyds List 10/9/91).

Privatisation schemes such as these will have lasting effects upon East European shipping companies and the maritime sector in general. For example they will mean increasing competition and higher levels of company investment, both leading to greater efficiency. Also the sale of industry will raise cash which might help to pay internal and external debts, perhaps reducing the current pressure placed upon the Polish people. Overall, the privatisation process has clear social, political, economic, technical, logistical, managerial, and legal implications, for the shipping industry in Poland, which have led to its selection as a main issue and one which implies that sustainable changes within shipping operations might follow.

Joint ventures

Joint ventures are an increasingly important form of co-operative arrangement. Virtually all sectors of industry now trade internationally and this has led to growing awareness of the advantages of commercial links between partaking countries who see joint ventures as a particularly practical business tool. The most important reason behind the launch of a joint venture is generally that neither party alone has at its disposal all the elements necessary to realise the objectives of the venture. Therefore each party relies on the other to supply the missing element or elements.

The possible motives for a joint venture are numerous. However they fall into five major categories - commercial, technological, legal, political and financial - and may include the following examples (ICC 1989):

1   to use complementary technology or research;
2   to raise capital;
3   to spread the risk of establishing an enterprise;
4   to achieve economies of scale;
5   to overcome entry barriers to domestic and international markets;
6   to acquire market power;
7   legal and/or political considerations.

The potential advantages of forming a joint venture relate to the realisation of these five motives and include; spreading risks, sharing fixed costs, capturing economies of scale, the pooling of knowledge and sharing of research efforts.

However there are also possible disadvantages of joint ventures such as their ability to eliminate or decrease actual competition between parent firms. Also it is possible that joint ventures may lead to the foreclosing of particular markets and may even reduce potential competition.

The negotiation phase of the venture involves the discussion and agreement of the main issues. These usually include the partners' individual and collective goals, the process of decision making, the measurement of progress, accounts, legal and fiscal regimes, share and measurement of inputs and outputs, and steps to be taken by partners during unforeseen circumstances. Negotiation must be thorough to ensure that problems do not arise in the operation phase. It has been found that it is especially important to resolve the underlying issues of power sharing, timescales for measuring success/failure and renegotiation, and to establish a clear 'divorce' clause (OECD 1986).

The potential problems which might arise during the operation phase of a joint venture are numerous and can include the following:

1   an unhappy marriage between unsuitable, inadequate or incapable partners
2   unclear objectives or hidden objectives by one partner
3   insufficient input of know how and management and insufficient attention to detail
4   under capitalisation and lack of financial support
5   faulty decision making mechanism

6   insufficient supervision and auditing
7   expansion of joint venture defeating other activities of partners and external activities of partners reducing their commitment to the joint venture
8   competition between partners may defeat the joint venture
9   unforeseen circumstances
10   dissolution of the venture
11   resolution of disputes
12   political changes in partners' national governments or political interference in the venture
13   devaluation and changes in exchange rates
14   changes in profits / investment repatriation rules
15   psychological problems - eg 50% share in venture deserving only 50% of attention
16   socio-economic pressures to maximise/minimise employment
(Moreby 13/12/90, Ramberg 1984)

In East Europe, economic reforms have led to recognition of the need for a restructuring of industry and for increased access to Western production technology and expertise. However, East European currencies are without exception non-convertible, and shortage of hard currency combined with a reluctance to rely too heavily on Western credits has limited the ability of Eastern Europe to make purchases from the West. East Europeans have sought to overcome these problems by opening up their markets to Western investment, by allowing Western companies to participate through joint ventures in new or existing enterprises based in East Europe. Most East European countries have allowed foreign investment for some time, but since 1986 the idea has taken wing with Poland, Hungary, the majority of states forming the ex-USSR, ex-Czechoslovakia and Bulgaria all introducing legislation allowing or modifying the conditions under which Western firms may take shares in certain enterprises and participate in their management. There are two exceptions: Romania has had joint venture legislation in force since 1971 but has seen no need to modify this since 1976 and the former GDR showed little interest in this area prior to German reunification (BOTB 1989).

In the initial euphoria following glasnost and perestroika, a spate of joint ventures calculated to revitalise moribund industries, were created in what was formerly the Soviet Union (Lloyds List 03/01/92). By the end of March 1991 3,200 joint ventures had been registered in the ex-Soviet Union, nearly three times as many as a year earlier. However, for a long time the number of joint ventures registered bore no relation to the number operating because, in the early stages, Western companies tended to register businesses without committing any money to them - by August 1991 the number actually operating was still only 948. But this also seemed to be changing. The volume of sales from joint ventures in the first quarter of 1991, adjusted for inflation was double that of the year before - at a time when Soviet output as a whole fell, on official estimates by 8% (The Economist 31/08/91).

Hungary adopted the joint venture stance from the Yugoslavs in 1973, acting as an example for the rest of the CMEA. Some 300 joint ventures were set up in the country, with a foreign complement of $300m, by January 1 1989, and another 330 with just $63m foreign complement in the first half of 1989, as the development of democracy in Hungarian politics got under way; yet economic uncertainties grew and may have reduced this growth as a result.

In Poland, 820 so-called Polonia firms were set up from the start of the 1980s by Polish expatriates with a foreign capital contribution of $130m. Some 51 joint ventures with a total capital of $220m were established from December 23, 1988 - when a much more attractive set of joint venture laws was released - until June 27, 1989. In fact there is little difference between the Polonia firms and the joint ventures except that the latter are believed to be better funded - though the foreign share in their capital has not been revealed.

Overall by late 1989 there were supposed to be about 2,000 joint ventures operational in the European ex-CMEA countries with more than a hundred in the transport sector and 41 in the shipping/ports sector. However many were not operational but in a state of suspended animation or just proposals (Seatrade Business Review 1989). Possibly the best publicised of those existing are those of the car manufacturers including Skoda (Czechoslovakia) and Volkswagen (Germany) (Economist 07/03/92), and General Motors (US) deal with the Polish government and FSO, the Polish state owned car market (Financial Times 01/03/92, Independent 29/02/92).

The importance of joint ventures to shipping is evidenced by the Caracas Conference back in 1981. The essence of the Caracas Declaration was:

1   international co-operation at business level;
2   transfer of maritime know how - technology - in its widest sense through for example joint ventures;
3   continuing and continual analysis of the elements for successful joint ventures, the 'dos' and 'don'ts';
4   re-examination of ship financing schemes in the light of changing circumstances.

In short, a form of practical co-operation between nations at individual business level based not on 'charity or government decree, but mutual self help and commercial principles' (Farthing 1985).

At another conference on joint ventures in 1984 Carden stated that 'one, and maybe the only method of reconciling the basic requirements of both the shipper and the shipowner is for the latter to combine all or part of his functions with those of one or more other owners similarly placed, and thereby offer jointly a

112

rationalised service capable of fulfilling shippers requirements in a way that no individual owner could do on his own' (Carden 1984).

Joint ventures in the liner trade may take any of the following forms (listed here with increasing degree of integration) and represent facets of the traditional conference and consortium system:

1   Rates agreement
2   Conference agreement
3   Joint scheduling
4   Slot/ space exchanges
5   Sharing ship costs
6   Sharing cargo and freight earnings
7   Consortium
8   Joint company

Each joint venture in any sector is unique and requires that it meets its own conditions for success. In the maritime sector these conditions could include:

-        mutual trust: understanding own and partner's reasons for the joint venture
-        political stability in partners' countries
-        international acceptability of legal regimes in partners' countries
-        adequate financial support
-        secure cargo flow / ship utilisation
-        properly structured, binding contract
-        ideally partners of equal size, age, financial strength, diversification
-        ideally similar organisational and managerial styles
-        partners same degree of commitment
-        maximising use of ships and equipment
-        similar rewards to partners representatives
-        actual or perceived benefits accruing to each partner
(Moreby, 1991)

Having noted the importance of joint ventures both to the countries of Eastern Europe and to shipping in general, we now turn to a combination of the two themes. In the maritime sector, setting up a joint venture has been one of the only ways for many financially stricken ex-Soviet shipping companies to slow down the shrinking in the size of their fleets (Lloyds List 11/10/91). The following are just a few examples of joint ventures involving shipping in the former Eastern bloc as a whole - indicating both their importance and wide usage in this sector:

1   Nippon Express Co. and Primuvatu Trans reached an agreement to serve as each other's agents. The agreement will allow Nippon Express to extend its door-to-door transport network to the Soviet Far East - the

113

ultimate aim of both firms being to build up a network capable of handling a door-to-door service between the two countries (Lloyds List 2/11/91).

2   Petromin and United Dutch Shipping Co. agreed to operate jointly six 65,000 dwt panamax bulk carriers in the international market. The vessels will be placed under the management of UK's Bolton Marine. United Dutch Shipping is to provide around £20m towards the upgrading of the vessels to bring them up to operational standard and make them competitive on the international market. The vessels were built in Romania between 1981 and 1989 and have been laid up recently awaiting finance to carry out repairs (Lloyds List 31/10/91).

3   Ex-Soviet Union and Greek ship owners, for example the Lavina Corporation which by the end of 1991 was managing more than 20 vessels under its joint ventures with ex-Soviet interests (Lloyds List 29/10/91, Lloyds Ship Manager 1991).

4   Romanian state owned line Navrom has been restructured as a joint venture company. Under its new name of Romline, the company is actively seeking joint ventures with western companies (Fairplay 12/7/90).

5   SeaLand Service Inc. has announced a series of co-operation ventures with Soviet transport authorities. One protocol called for the improvement of the Trans-Siberian Railways. The land bridge would link the USSR with Europe, Asia and North American trade routes, increase utilisation of ex-Soviet ports and improve distribution within the former Soviet Union (Fairplay 31/05/90).

6   Finnish ABB Stromberg Drives and subdivisions of the ex-USSR Merchant Marine Ministry, the Ministry of Shipbuilding and the Ministry of Electrical and Electronics Industries signed an agreement to cover the manufacturing of marine automation systems by a joint venture company based in Leningrad. The object of the company will be to supply state-of-the-art automation systems to merchant vessels under construction for CIS owners, as well as organise service activities for these systems. ABB Stromberg Drives will have a 30% holding in the new Company (Fairplay 15/03/90).

7   One of shipping's largest joint ventures, involving three of the leading forces in bulk shipping, has been finalised between Russian, Chinese and Norwegian interests. They are to operate 10 new, high specification combination carriers - ordered from South Korea's Hyundai facility at around $600m - in a pool to be managed from London. AKP Sovcomflot, which ordered the ships for Liberian nominee Fiona Trust and Holding Corporation, is to sell an unspecified number of the 96,000 dwt ore/bulk/oilers to China National Chemical Import and Export Corporation (Sinochem) and to Kristian Gerhard Jebsen Skipsrederi of Bergen. All 10 ships will go into the pool. The three major names will be 33% shareholders in the pool company, to be known as SKS Obo Ltd,

which will trade the vessels under the Liberian flag (Lloyds List 06/03/92).

8   Jurinflot International, set up in August 1991 whose partners are the Anglo-shipping Co (now Morline), Sovfracht Moscow and the UK and west of England P&I Associations (Lloyds List 03/01/92).

9   A joint venture tanker owning company was established by Norway's Anders Wilhelmsen group and the Polish Steamship company. The new venture, called Steamco, planned to acquire or charter a number of tankers in the 80,000 - 100,000 dwt range for long term trading purposes, with commercial management provided by Wilhelmsen and crewing and technical management by Polish Steamship (Lloyds List 10/07/93).

Although there are surprisingly few taking place in our case study area of Poland as yet, the potential of joint ventures such as the examples listed above are likely to have numerous lasting effects upon the shipping industries of all of Eastern Europe. For example, they may provide some of the capital necessary to revitalise the maritime sector, as well as introducing advanced technology, and more modern production and management techniques by Western standards. The clearly significant political, economic, technical, managerial, and legal implications of potential joint ventures in the shipping sector have led to their selection as a major issue.

Technical condition

The term 'technical condition' refers to the state of Poland's ships, ports, inland transport systems, communications and other infrastructure. As the state has begun to withdraw from industry it has left behind technical standards which generally are poor and inadequate by western measurements, for example:

> In 1990 the average age of Polish vessels was fairly high and increasing, at around 13 years (Ernst & Young 1990)
> The standard of inland rail and road transport infrastructure is generally poor due to under investment, with little government finance available to make improvements (Kolankiewicz and Lewis 1988). The situation has improved only slightly since 1982, when there were 27,157 km of railway, of which just 7,410 were electrified, and there were 254,037 km of road, of which only 15,000 had hard surfaces (East European Economic Handbook 1985). Persson and Backman (1993) support the view that there is a need for investment in new roads, railways and environmental protection. They point to the shortage of highways surrounding the larger cities and the increasing frequency of traffic congestion. They continue:

> There is normally no information about repair works being carried out

on roads or bridges which can make it difficult to foresee transport delivery times. Parts of the two important motorways in Poland (the West-East which is the shortest route from West Europe to Russia, and the North South route from Scandinavia to Southern Europe) have already been constructed but await foreign capital before they can be completed.

Rail transport is slow and expensive. The system for cargo transportation is inefficient. The number of employees is extremely high reflecting labour intensive practices and burdensome decision practices. The changes needed have to involve the adoption of technologies introduced in the West some 20 years ago, such as container traffic and combined transport, and they have to involve the simplification of frontier crossings.

Crossing borders represents a bottleneck, particularly between Germany and Poland. There are physical lines of waiting trucks and it is difficult to predict how long an individual truck will spend at the border. The major reasons for these are inadequate border facilities and complicated and rapidly changing custom formalities'.

Information technology availability - both soft and hardware - and the extent of related knowledge is generally low compared to the more developed countries of western Europe. Very few companies use computerised systems and if they do it is mainly for accounting purposes. They do not have computer-based production planning and control systems, nor do they have computer based inventory management systems (Persson and Backman 1993).

Communications are poor at present; for example national telephone calls can take up to several hours to connect. Poland has the fewest number of telephones per head of population of any European country, with 1.2 million subscribers waiting to be connected to the telephone system. Much of the central switching equipment is old and needs replacing.

The banking system is another problem. All money operations are time consuming. Any transfer of money takes several weeks, and can vary greatly from one operation to the next. The banks are hampered by much manual control and handling of documents.

Meanwhile, Poland has a large proportion of ageing industrial plant and unfinished infrastructure projects, the replacement and completion of which needs to become an economic priority.

The condition of technology dictates how shipping companies may proceed and the speed at which they can develop and progress. Its inadequacies therefore present them with a number of problems.

In view of its substantial external debt, the Polish government has little finance available for investment in infrastructure. In order to compete with western companies Polish companies must first invest in infrastructure and technology; for example they need new ships and equipment before they can provide a service which is efficient enough to make some profit. However, as mentioned previously there is little capital available in Poland, especially as government subsidies - both direct and indirect -are withdrawn. Perhaps the most feasible method of investment would be to attract capital from the west, for example through joint ventures. However, western enterprises are unlikely to want to invest in Polish companies with poor technological standards, especially shipping companies which are highly capital intensive.

The problems of poor technical condition faced by the Polish shipping industry have been further compounded by the economic, political and social changes which follow:

1   The decline of the taxation base
2   The removal of guaranteed cargo - Polish shipping is no longer guaranteed Polish state cargo and that of other ex-CMEA countries, which has meant a fall in capacities carried, and consequently, less profits available for reinvestment.
3   The general Polish depression has led to a fall in the demand for goods and therefore the derived demand for shipping, restricting the inflow of funds. Also, the depression has meant that there is less income tax available to the government for reinvestment.
4   The increase in competition, both real and potential, due to freer markets has decreased the amount of capital available for investment.
5   The removal of subsidies from the shipbuilding industries has increased the price of new buildings, which consequently are even less affordable.
6   The shipping industry has been exposed to a free market situation with equipment of low technical quality inherited from the old regime along with its market failures.
7   The removal of protection from bankruptcy has meant that investments have to be made wisely. This whole situation is new to Polish shipping, and will be difficult and timely to adjust to.

The technical problem has ramifications for shipping company marketing policy. Overall the problems are likely to lead to

1   Industrial inefficiency
2   Increasing prices and falling demand
3   Changes to routes, modes, and companies used
4   Day to day logistical problems
5   Companies seeking joint ventures as a solution to the problems faced.

The general position as regards technology is one of a paradox - Polish companies need to invest in infrastructure before they can compete and become profitable,

117

but they need to earn the money before they are able to invest. The best chance of improving infrastructure is to obtain western aid, or to barter with western companies, possibly offering services in return for investment.

Technical condition clearly has significant economic, managerial, and legal implications which have led to its selection as a major issue. The technical condition of Poland's companies will help to determine the position that they occupy in the market place. For example, as improvements look unlikely to occur in the near future, shipping companies may only be able to offer low quality service, counteracted by charging low prices.

Democracy and labour issues

Most discussions on the future of Poland have centred upon the issues of democracy and capitalism. This is because of the commitment of new politicians precisely to these twin goals, and in some instances it may reflect the conviction that the transition of Poland will entail a democratic capitalist outcome (Millard 1994).

Poland has a number of characteristics which seem beneficial to bringing about or supporting a democratic policy. First, the Poles have a strong sense of national community and common culture. Second, those who took power from the communists moved immediately to alter the constitutional framework, eliminate prior censorship and the agencies of communist control of the media, dismantle the secret police, hold free elections and re-establish property rights - all necessary mechanisms of democratic development. The final factor facilitating democratic development is that of geographical location. Western Europe has economic, political and strategic interests in the emergence of a stable, democratic and prosperous Eastern Europe. The European Community in particular is formally committed to assisting change, and the post communist government of Poland immediately expressed a desire to join the Community (Millard 1994).

The introduction of democracy will have implications at every level. The following indicate areas where the most notable changes are likely to take place:

1   Employer/ employee relationships.
     It is likely that democracy will mean a move away from central control towards more western style management. Instead of directives automatically coming downwards from the top, there is likely to be scope for ideas generated at lower levels, in line with western style relationships. There is also likely to be corresponding increasing emphasis on the importance of the customer.
2   Unions.
     Poland has already witnessed a general decrease in the power of unions, and falling levels of membership, as people reject that which reminds

118

them, possibly unfairly, of the old system.

3   Votes/voting

There are clear relationships between political will and voting power.

4   Government stability.

The introduction of a new democratic system is likely to lead to a decrease in government stability, coupled with emerging divisions of parties.

5   Central government direction

The removal of central government direction will have implications such as the likely reduction of guaranteed cargo, removal of subsidy and problems of raising investment.

6   Vote winning issues.

Democracy may also bring about a government that concentrates upon vote winning issues such as health and housing, causing problems for competing industries such as shipping.

7   Organisational changes.

Examples here include the likely fall in union membership and power, and changing managerial roles

8   Market forces.

The market will become subject to the forces of supply and demand in line with those of the West.

9   Free choice.

Democracy will increasingly mean that companies are free to chose modes, sources, exporters, ports, and contracts. This in turn will mean less stability and no guarantees.

Linked to these issues, Persson and Backman point out that to further their economic development, East European companies must become more competitive, and to do this they need some critical management skills. Generally firms lack the experience and the systems to handle a geographically dispersed organisation or a multi warehouse system and western firms looking into joint ventures find that the biggest challenge is management know-how, skills and attitudes. Topics such as logistics were not on the agenda at academic institutions or in companies. They report that a majority of managers have been trained in an era in which logistics was useless. For example many companies built up inventory extensively as this reflected an 'efficient' and reliable enterprise. It was against the basic rule of western logistic practice but it was wise (Persson and Backman 1993).

Overall, democracy in Poland is likely to develop in some form, but much will depend upon the 'educative capacity of those in positions of power, their skill and their probity' (Millard 1994).

In summary, the main social, political and economic changes affecting Polish shipping have been established as:

Privatisation
Joint ventures
Technical condition
Democracy and labour issues

Company adaptation to these major issues will constitute in marketing terms, a repositioning exercise. In the next chapter, we will go on to examine the concept of repositioning in greater detail.

6 Polish Ocean Lines

The main issues which have emerged from the social, political and economic changes occurring in East Europe, and which had an impact upon the Polish shipping industry, have been identified as privatisation/state withdrawal, joint ventures, technical condition, and democracy and labour issues. These changes can be viewed as helping to create a series of substantially new market environments for the operation of Polish shipping, which is likely to force the industry to undertake a process of repositioning, if it is to survive. However, before the concept of repositioning can be examined in any detail it is first necessary to narrow the broad research area further.

It was discussed in chapter three that the research needed narrowing to one sector (the liner industry) and hence one major company - Polish Ocean Lines. POL's involvement in the liner shipping industry would still provide an unmanageable study area, and so for further simplification the research has been narrowed to the North Atlantic trade. This trade was chosen because it was one of the most important of Polish Ocean Lines' operations. Cross trade has played a crucial role in POL operations and its share reached its highest level in 1985-6. Since then the share has declined, coinciding with the expanding servicing of Polish import and export flows and transit connections (ie trade through Poland). In 1988 an analysis of cargo carried on the major routes and types of trade revealed that the highest percentage of all POL's cross trade - 39.2% - was being carried on the North America Line, the second highest being 12.9 % on the Australia Line (Ernst and Young 1990). The North Atlantic service is a major global trade and it is an area of considerable interest for European Community operators, for example P&O Containers Ltd, Hapag Lloyd AG, Maersk and Nedlloyd Lines. TEU figures for 1990 and estimates for 1997 indicate the size and importance of the trade for all operators (Drewry Shipping Consultants Report 1993):

	1990	1997
Cargo demand (westbound)	1,258,000	1,600,000
Cargo demand (eastbound)	1,233,000	1,300,000
Capacity	1,944,000	2,177,000

Table 11 shows the relevant capacities for each of the years between 1990 and 1997 and the corresponding figures for demand which enable percentage utilisation figures to be calculated. Notably, in 1990 westbound capacity utilisation was only 64.7% whilst eastbound was 64.2%. By 1997 Drewry estimate that westbound utilisation figures will rise to 73.5% whilst eastbound the figure is expected to fall to 58.8%.

Table 11
Capacity, demand and utilisation on the North Atlantic 1990 - 1997

Year	Capacity (M TEU)		Demand (M TEU)		Utilisation (%)	
	w/b	e/b	w/b	e/b	w/b	e/b
1990	1.944	1.922	1.258	1.233	64.7	64.2
1991	1.914	1.930	1.150	1.334	60.1	69.1
1992	1.917	1.900	1.200	1.375	62.6	72.4
1993	1.856	1.839	1.150	1.400	62.0	76.1
1994	1.877	1.860	1.300	1.400	69.3	75.3
1995	2.027	2.010	1.400	1.400	69.1	70.0
1996	2.177	2.160	1.500	1.350	68.9	62.5
1997	2.177	2.210	1.600	1.300	73.5	58.8

Source: Drewry 1993

The specific trade area is defined as being between North West Europe and the east coast of the United States of America from Boston to Cape Hatteras - a widely used definition of an identifiable and distinct market - hence it does not include Canada or the Gulf coast. Transhipments, such as via Canada or Icelandic ports, have also been excluded.

Eleven of the North Atlantic operators belonged to the Trans Atlantic Agreement (TAA), and to understand Poland's activities in this market at the end of 1992/ beginning of 1993, it is necessary to assess the TAA agreement itself. This agreement was entered into on 15 April 1992 and when the US Federal Maritime

Commission (FMC) approved a rewritten draft of the TAA in August of that year it was established immediately. The TAA was preceded by a web of agreements including Agreement 1237, the North Europe - USA Rate Agreement (NEUSARA) the USA-North Europe Rate Agreement (USNERA), Gulfway Agreement and Eurocorde I and Eurocorde Discussion Agreements. According to the minutes of a meeting on 8th October 1992 described as a 'Gulfway meeting', the TAA's members are divided into two groups as follow (TAA minutes of Gulfway meeting, 8th October 1992):

Structured members

This group includes lines previously members of various conferences on the trade, these being ACL, Hapag Lloyd, P&O, Nedlloyd, Sea Land, Maersk and OOCL. NOL and NYK have entered the trade as structured members of the TAA, as part of a tri-continent service, but only in association with Hapag Lloyd on a vessel sharing basis.

Unstructured members

This second group, similar to the previous tolerated outsiders, consists of Mediterranean Shipping, POL, Cho-Yang, DSR and Senator. The latter two have worked closely together during the period of the research, recently (1994) merging to form one company.

One of the differences between the two groups is that the structured members form a rate committee with voluntary attendance from the unstructured group. The aim was that all lines would be in one group as from 1 January 1994, although this failed to materialise, partly due to continuing problems over relations between TAA operators and the European Community.

TAA members collectively have a market share of over 83% (1991), whilst the remaining non-TAA members offer little or no effective competition. Of the few remaining non-TAA carriers, Evergreen has an 8.6% Westbound market share (Westbound market shares on TAA Trade and Eurocorde market shares 1991). Of the other non-TAA shipping companies, Lykes Lines (market share 1.7% approximately) and Atlanticargo (with a share of less than 2%), ABC and ICL (minimal share), are of insufficient size to present any real effective price or service competition to the TAA.

The aims of the TAA were outlined by Karl Heinz Sager, Head of Senator as (Lloyds List 23/6/92):

-       providing the incentive for carriers to manage individual vessel capacities

123

in such a way as to ensure there is always space available for all containers in the trade to be shipped with monitoring of carryings carried out on a monthly basis;
- greater degree of flexibility in pricing;
- members exchanging slots on their vessels;

The principal mechanism of the TAA was a capacity management programme, which requires that the signatories keep an agreed percentage of their slots unused on the weaker westbound leg of the market place. The TAA tariff introduced on 1st January 1993 represents a radical restructuring of the former conference ones, with a far smaller number of categories; yet more base ports have been eliminated; and 'multifactor' rates, which split the through rate on door to door shipments into its separate components have been introduced. Pre TAA, companies operating on the North Atlantic were generally offering low rates for competitive purposes, and were mainly loss making; these losses could not be endured for any long time period so that prices had to rise. The TAA increases were considerable, with rates rising well above those for 1992 for both service contracts and tariffs, but the lines argue that when compared to 1989 they are lower overall. The TAA has realigned service contracts on a January to December basis, and stipulated a minimum of 250 TEU for the period, compared with a more usual 100 TEU.

The TAA established the following objectives, amongst others, in support of the creation of a new tariff:

- General user friendliness - leading to less errors in quoting rates/ documentation
- Simplified rules and regulations
- Port/ port rate transparency
- Harmonisation of the present rating structures amongst TAA parties
- To support joint rate setting by TAA parties who wish to act in that manner
- Reduced costs in administration

TAA members have also reduced capacity on the trade in a recent rationalisation programme. The withdrawal of four ships by OOCL following its tie up with the VSA, and the merger of Hapag Lloyd's Pacific coast service into the new 'pendulum operation' with NYK and NOL have removed 2500 TEU a week. This has decreased available TAA capacity by 10-11% to 21,000 TEU a week. Westbound capacity has been reduced by TAA lines agreeing not to use 75% of idle slots for commercial cargo, cutting westbound overcapacity from 30% to 20%, and westbound capacity to approximately 18,000 TEU (Seatrade April 1993).

Rationalisation attempts are complicated by TAA parties who operate round the world (RTW) or pendulum services. Cho Yang/DSR/Senator operate one eastbound and one westbound round the world service; whilst Maersk operates a pendulum service between the Far East/ North America/ North Europe. The operation of such services limits the ability of those carriers to rationalise by entering into arrangements with other carriers providing 'end-end' services. The operators of RTW and pendulum services will determine capacity deployment by reference to the heaviest leg, namely either the Europe/Far East (RTW) or the Trans-Pacific (pendulum). Such services therefore bring more capacity onto the Transatlantic leg than is required by the carriers on that leg.

The objectives of the TAA envisage horizontal price fixing and market sharing, and the agreement generated a complaint to the EC Commission by European and British Shippers Councils (BSC) on the grounds that it infringes Articles 85(1) and 86 of the EEC Treaty in so far as the TAA and the conduct of its members:

1   limit, control or share the supply of sea, inland transport and other services and markets, technical development and investment to the prejudice of shippers;

2   directly or indirectly fix, and unfairly impose, transport and other rates and conditions or any other trading conditions in respect of such services;

3   apply dissimilar conditions to equivalent transactions with other trading parties thereby placing them at a competitive disadvantage;

4   in so far as the TAA and its members' conduct infringe Article 85(1), there is no automatic exemption available under Council Regulation 4056/86 as correctly interpreted;

5   in so far as the TAA and its members conduct infringe Article 85(1), the agreement may not qualify for individual exemption under Article 85(3), in that none of the four conditions for exemption are satisfied;

6   in so far as the TAA and its members conduct infringe Article 86, there is no exemption available under Article 86 or Article 8 of Regulation 4056/86.

It is also argued by the BSC that the TAA capacity management programme will not increase efficiency or provide any cost benefits. In particular, existing and planned vessel sharing or pooling arrangements as described earlier provide a better opportunity of genuine rationalisation of resources and cost savings provided tonnage overall is reduced.

It remains to be seen whether the TAA will be accepted by the EC's Competition Directorate DGIV, and whether the FMC will maintain its current position. The BSC have requested that the EC Commission:

-    adopt interim measures under Article 10 and 11 of Regulation 4056/86, and any other appropriate regulation, requiring the TAA members to cease all tariff rate fixing and to return to or maintain the tariff rate or the service contract rates and conditions operated by each TAA member in relation

125

to all BSC shipper members in 1992 until the date of final decision by the Commission
-       require the TAA members to terminate the TAA agreement pursuant to Article 11(1) of Regulation 4056/86 on the ground that the TAA agreement infringes Articles 85(1) and 86 of the EEC Treaty and to refuse any individual exemption under Article 85(3)
-       alternatively, to withdraw any block exemption which may be enjoyed by the TAA agreement pursuant to Article 7 and 11(2) of Regulation 4056/86
-       immediately to inform the TAA members that after preliminary examination the Commission is of the opinion that Article 85(1) of the Treaty applies and that application of article 85(3) is not justified with the consent withdrawal of immunity from fines accompanying notification, pursuant to Article 19(4) of Regulation 4056/86.

One of the changes which may occur should the TAA's higher rates remain is that those global carriers which have expressed the intention of entering the Atlantic trades but have not yet done so largely due to the rates situation, would finally be tempted in - reducing the TAA share to more acceptable proportions. An alternative viewpoint is that the collective dominance of the TAA in the North Atlantic market may deter potential new market entrants who could increase available capacity. The fact remains that TAA lines are willing to sacrifice market share to maintain higher rates.

Having established the broad background to the North Atlantic trade we can now look specifically at details of POL's operations. As discussed in chapter three, the comparison of past and present operations of POL is an exercise in assessing repositioning - a theory supported by Trout and Ries (1986) and Shostack (1987). The use of a framework enables this exercise to be carried out within the structure provided by the 7 P's of service marketing. As explained earlier it is a display technique which has been used successfully in situations where more quantitative techniques have proven to be inappropriate (for example, UK Department of Transport 1977, 1979, Wind and Robinson 1972). The use of qualitative research is further supported by Walker (1985) as well as Chadwick, Bahr and Albrecht (1988). Additionally, the application of a display technique maintains consistency throughout the research, as the contextual model used earlier was also of this type.

The framework will provide a year on year comparison of POL's operations within the context of the North Atlantic market. As described in chapter three most of the information within the framework has been gained from interviews with top level professionals at POL in Gdynia and London - an example of the structured interview can be found in appendix 2. The framework (figure 4) shows the relative positions of POL in 1988 and late 1992/ early 1993 on the North Atlantic trade in terms of the 7 P's. These are broken down into their respective elements derived from the work of Booms and Bitner (see figure 1);
- selected to represent measures of the seven P's in the context of liner shipping

P	Elements	Measurement method	Polish Ocean Lines	
			1988	1992
P R I C E	Rebates	Yes/No	Yes	No
		Deferred Yes/No	Yes	No
		Percentage	Negotiable	N/A
		Other details	Rebates offered on commercial basis	TAA membership prohibits rebates
	Discounts	Yes/No	Yes	Generally No
		Details	No details due to commercial sensitivity	TAA membership prohibits discounts
	Price for transporting one TEU	US$	According to market	1035
	Market segmentation	Yes/No	Yes	Yes
		Details	On a market basis - by commodity/ product/ season and notably by nationality	Segmentation as allowable within the TAA confines
	Subsidy	Yes/No	Yes	Yes
		Amount (zloties)	Unknown	Unknown
		Type	Direct and indirect including new buildings and land	Still some indirect subsidy - State writing off annual cash debt
	Currency convertibility	Yes / No / Partial	No	Partial
		Acceptable currencies	CMEA not charged, west charged in convertible currency	All convertible currency and zloties
	State control	Yes / No	Yes	Yes
		Details	Total state control eg. staff, vessels, prices	State owned but greater independence, joint stocked into separate companies for privatisation

Figure 4 Polish Ocean Lines' framework

				USS	Unknown	75 million loss
PRICE	Profit / Loss	Details			No accounting system to provide details	Only Pol Levant profitable, cross subsidising other areas
	Competition	Details			None within Poland as none allowed, much from abroad	Competition from within Poland allowed but none yet established
PROMOTION	Sponsorship and hospitality	Details			Limited examples include dining with important customers	Further reduced due to financial limitations
	Advertising	Yes / No			Yes	Yes
		Budget			None as such, as state financed	Company responsibility, amount reduced to 10% of 1988 figure
		Press used	General		Yes	No
			Shipping		Yes	Yes
		Frequency			Occasional - particularly related to special reports	Occasional - particularly related to special reports
		Type	Information		Some, for example contact numbers	Some
			Persuasion		None	None
			Presence		Very important	Very important
	Image	Corporate image Yes / No			Yes	Yes
		Details			Old style in line with other Polish shipping companies due to state control	Opportunity to change but newly joint stocked companies retaining old image
	Representatives	Yes / No			Yes	Yes
		Number			Unavailable	37
		Locations			Worldwide	Worldwide - all five continents
		Agents	Yes / No		Yes	Yes
			Number		Unavailable	247
			Locations		Worldwide	Worldwide

Figure 4 Polish Ocean Lines' framework continued

128

Market Research	Yes / No		Yes - in collaboration with Gdansk University	Yes
	Budget		Amount unknown as state funded	Reduced form 1988 figure
	Number employed in marketing department		Unknown	Reduced to 3
Public relations (PR)	Budget		Amount unknown as state funded	Reduced form 1988 figure
	Promotional items	Types	Extensive range of items	Reduced to clocks, wallets, ties and bags
		Numbers	Numerous	Limited, rationalised since 1988
Directories	Details		All major worldwide shipping directories including Lloyds, Fairplay, Containerisation International etc	All major worldwide shipping directories including Lloyds, Fairplay, Containerisation International etc
Literature	Yes / No		Yes	Yes
	Types	Brochures	Yes	Yes
		Booklets	Yes	Yes
	Destinations		Worldwide	Worldwide
	Languages		English	English
Agencies	Yes / No		No	No
	Types	PR	N/A	N/A
		Advertising	N/A	N/A
	Budget		N/A	N/A
Exhibitioneering	Yes / No		Yes	Suspended
	Numbers		Unavailable	N/A
	Locations / Details		International as appropriate	N/A

Figure 4 Polish Ocean Lines' framework

PROM.	Press releases	Yes/No	Yes	Yes
		Numbers	Variable	Variable
		Languages	Polish/English	Polish/English
PHYSICAL	Containers	Numbers owned and leased	14,260 owned, 22,114 leased	Approximately 36,000 owned, 24,000 leased
		Numbers in own colours	All owned containers ie 14,260	All owned containers ie 36,000
	Vehicles	Number / type of lorries	East and west manufactured. Figure uncertain.	120 Volvos
		Number / type of others	Limited car and van fleet	Limited car and van fleet
	Company offices	Number	1	2
		Details	Head Quarters in Gdynia, plus various depots and warehouses	Head Quarters in Gdynia and Euro Africa office in Szczecin, plus various depots and warehouses
EVIDENCE	Ports served	Number served	Over 100	Approximately 100
		Locations	Worldwide including obligatory CMEA ports	Worldwide, so longer necessarily serving CMEA ports
	Routes served regularly	Number and details	Eight categories - America, Australia, Europe, West Africa, Asia, South Pacific, Mediterranean, Red Sea	Eleven routes - North Atlantic ConRo service, US Gulf South Atlantic Container service, South America West Coast Line, Central America, Caribbean and US Gulf line, East Coast and South Africa Line, Mediterranean and Service-Mare Line, South America East Coast Line, Asia JOS Service, South Asia Indonesia/Vietnam Line, South Asia Bengal Bay Line Australia Line
PROCESS	Conference and Consortia	Membership Yes/No	Yes	Yes
		Title	One Far East and two South American	TAA only, other memberships dropped
		Level of Membership	Full	Unstructured member
	Guaranteed traffic	Yes/No	Yes	Yes
		Details	Substantial CMEA and State guaranteed traffic	Considerably reduced amount of state guaranteed traffic and loss of CMEA market
	Collaboration	with road	Own trucking company	Own trucking company
		with rail	Yes - state controlled	Yes - traditional links

Figure 4 Polish Ocean Lines' framework - continued

130

P L A C E			
Origins and destinations served	Numbers	Approximately 100 ports, 17 liner services	Approximately 100 ports, 21 liner services
	Details	Worldwide basis, state controlled hence serving CMEA countries and associated ports	Worldwide basis, wider range of liner services, reduction in number of ex-CMEA ports served
Computer Reservation Systems	Yes/No	Limited	Yes
	Details	Very limited usage	Used by majority of representatives and agents
Qualifications and training	Details	Seafarers required to have Polish government controlled seafaring qualifications, management usually ex-seafarers	Seafarers required to have Polish government controlled seafaring qualifications. Introduction of a range of in house and out house training courses eg. management
P E O P L E			
Employees	Number at sea	6,500	4,500
	Number ashore	1,500	1,300
	Total	8,000	5,800
Management Structures	Details	State managed through an overall state appointed board	Control held by company board
Unions/Workers Councils	Number of members	Majority of workers	Approximately 50% of workers
	Details	Mainly Solidarity and OPZZ, with some members of Polish Officers and Crew Members Unions - important as they formed a significant element of the management process	Unions of declining importance as market forces take over, popularity decreased due to association with old regime

Figure 4 Polish Ocean Lines' framework - continued

Speed of North Atlantic service	Vessel Speed (kts)	21.7	21.65
	Days (Gdynia to New York)	12	10
P Safety	Containers lost p.a.	No data available	No data available
R Reliability	Number of late deliveries p.a.	No data available	No data available
O Frequency of North Atlantic service	Number of services per week	1	1
D Quantity	TEU capacity on North Atlantic	4 vessels of 1417 TEU	4 vessels of 1426 TEU
U Ship types	Number of each type (see appendix 1 for abbreviations)	DC (40), DN (37), WA (15), DR (4), RF (2), PM (2), PN (1), CL (3), PU (1), WL (2)	DC (11), DN (2), WA (6), PM (2), CL (1), WL (1)
C Ship Ages	Average years	13.46	8.65
T Flag	Details	Polish	Majority Polish, six vessels flagged out to Cyprus
Cross trades	Yes/No	Yes	Yes
	Percentage of containers	Less than 45%	Over 45%
	Details	Including CMEA routes	Including fewer CMEA routes Rising to compensate for decreasing domestic trade
Computer tracking for containers	Details	No	Yes
Trading terms	Details	All exports sent CIF all imports FOB - state enforced	State enforcement removed and operations now on more commercial basis

Figure 4 Polish Ocean Lines' framework

operations and specific to that market area.

Price

1. Rebates

Rebates constitute a major element of liner shipping pricing structures and are usually offered as a sales incentive or as a method of gaining customer loyalty. They may be deferred and are usually calculated as a percentage of the total price as in this research. POL did offer negotiable deferred rebates in 1988. This is because rebates are standard practice in the liner shipping industry and POL had to offer them in the same format as competitors in order to compete in an international market. However, on the North Atlantic trade, the formation of the Trans Atlantic Agreement now prohibits their existence, except in the rare occasion of 'independent action'. In terms of position there has been a noticeable change where rebates are concerned - but one forced upon POL as a member of the TAA - and not one reflecting company positional strategy as such. However, membership of the TAA itself was a positional choice, reflecting a company desire to become part of the established shipping community in that market.

2. Discounts

Discounts are frequently offered to shippers who use the service regularly or who need to transport large quantities. They are usually calculated as a percentage on orders exceeding a set level. Although it is known that POL offered discounts on a negotiable basis in the past, exact figures are not available as they are regarded as commercially sensitive information. It may be presumed that they were offered in the same format as competitors' discounts. In 1993 POL was generally unable to offer discounts on the North Atlantic trade as the TAA prohibited them. The exception occured where a shipper solely using POL wished to switch to a company independent of the TAA. This independent action was permissable at director level only. This presents a second recognisable change in terms of position, although once again a change driven by TAA membership and a desire to be part of the established market.

3. Price for transporting one TEU

This measurement provides basic information on price. It is estimated as the average cost of transporting one standard TEU from North America to North Europe. In 1988 POL was only able to charge around $940 per TEU (Conference statistics - BSC 1993). By 1993 this price had risen to $1035 which was the fixed TAA price for an unstructured member. This represents a significant change for

POL's position. In 1988 they were competing mainly on price, but by 1993 the price was standardised to two preset levels for all TAA members, placing POL in the same position on price as the majority of other operators on the North Atlantic and in competition with non members (eg. Evergreen) who were pricing noticeably lower. Overall price change, however, was small (an increase of some US $95) reflecting a depressed and oversupplied market, struggling to maintain price levels; rather than a deliberate intention to undercut others, or keep prices low.

4. Market segmentation

Market segmentation within liner shipping operations has become increasingly relevant to pricing policies in recent years. Bases for market segmentation include mainly commodity, customer, season and quality of service. POL did segment the market in the past but this was on the basis of commodity, and more significantly, nationality. The market was segmented into Polish and foreign custom and prices offered would have been lower for Polish customers and those from other CMEA countries. The setting up of a free market in Poland has led to market segmentation on a much wider basis, with the market divided into an increasing number of sections. As experience of the free market increases, segmentation generally is likely to become more extensive. On the North Atlantic trade however, this will have to take place within the guidelines of the TAA. Generally, the degree and basis of market segmentation has changed significantly between 1988 and 1993, reflecting an internal company desire to compete on the free market, and an external pressure to survive financially and commercially within the context of TAA membership .

5. Subsidy

Subsidy was and remains a major issue for all Eastern bloc shipping. In the past shipping was very heavily subsidised, although the exact amount of subsidy to POL is not known and is unlikely ever to be estimated accurately. Subsidies were direct and indirect, including for example new buildings, power and land. By 1993 cash subsidies still existed, provided through the Ministry of Shipping, but to a much lesser extent, and the intention was that there should be a gradual move towards zero subsidy to create a free market. Some indirect subsidy also remained, particularly in the form of charges for land, buildings and other capital equipment.

Subsidy is also directly related to the labour issue. In POL labour must still be retained at centrally specified levels as the state will not allow full and free redundancies, but the finance for these uneconomic policies must be found from within the company.

The reduction of subsidy from POL necessitates repositioning of products within the market. The company will not only be operating on a new basis but will have to increase profits, perhaps by raising prices, in order to fund new buildings and other previously subsidised items. This repositioning is likely to include expansion into new markets, and withdrawal from others. The financial inadequacies of operating on the North Atlantic in recent years are clearly relevant here, as are the pricing constraints imposed by the TAA.

6. Currency convertibility

Currency convertibility is another major issue for East European shipping which can and did act as a hard currency earner for countries with non convertible currencies. In the past the Polish zloty was not convertible in any way and Westerners had to pay POL in conventional convertible currencies, whilst for CMEA countries there was no charge as such, as they operated a form of barter system or payment was in transferable roubles. By 1993 the zloty was partially convertible and POL would accept zlotys as well as all convertible currency. This has some effect on company positioning in the market as it affects the emphasis within POL's operations. In the past one of POL's main concerns was to earn hard currency and consequently the company concentrated on the US$ dominated cross trades. In 1993 POL found that they still needed to concentrate upon these trades but for different reasons. Although there was less need to earn hard currency the cross trades provided the greatest opportunities for generating profits, especially as trade to the former CMEA countries declined steadily. Hence the company had repositioned in its attitude to cross trading and consequently its approach to the same market also has needed to change.

7. State control

State control is an important element in this context as the price charged may be fixed or controlled to some extent by the state. In 1988 state control of POL was total, affecting all areas of operation including price setting. In late 1992/ early 1993 the company was still state owned and so controlled to some extent. However, it had been joint stocked into separate companies ready for privatisation. Hence, there was greater independence; for example POL was now responsible for prices, income, staff, and vessels. This left POL free to determine the locations, number, frequency and share of markets served and quality of service. However on the North Atlantic trade POL was now restricted on price by the TAA. In terms of positioning POL has moved away from state control on prices charged on the North Atlantic to TAA control. Although POL is still restricted, it is suggested that the TAA is better placed to make commercial decisions regarding price on this service, and POL feel that the only way to raise price to a profitable level is to join forces with other companies operating on the service, as part of the TAA.

8. Profits

Prices may be charged with a profit level in mind, or a profit margin may be added to the cost of the service in order to calculate the price charged. Price charged has a major effect on profit levels and there must be a balance between charging a low price to generate high sales and charging higher prices for fewer sales. Past profits for POL's activities on the North Atlantic cannot be established as figures remain unknown with no accounting system to provide the details. In 1993 POL would give no exact profit figure but did admit that profits were poor. The Coastal Times (16/10/94) revealed POL's recent overall losses as follow:

Year	1990	1991	1992
Loss (mln US$)	6.2	62.2	75.0

The figures indicate severe increases in loss and represent heavy loss over a relatively short time period. In April 1993 the company was placed in the hands of a Commissioner to save it from bankruptcy. The losses were partly due to removal of subsidies, probably partly to company inefficiencies and partly to poor returns in the liner market at that time. In 1992, after POL was joint stocked into separate companies ready for privatisation, only Pol Levant - serving the Mediterranean - made any profit and this was used to subsidise the other sections. If POL is to survive in the free market it will have to improve profits substantially and this may require repositioning on price which will also have to maintain competitive levels. The 1988 position of POL was to charge low prices in order to gain market share. In future they may need to raise price in order to improve the quality of service and use the profits to replace ships in order to stay competitive. However, on the North Atlantic trade, price is largely controlled by the TAA and profits generated on this service would be improved or even sustained only by keeping costs down.

9. Competition

Competition plays a large part in pricing. Setting prices with regards to competitors' prices for any particular service is a form of positioning in the market place. In the past there was no competition to POL from within Poland and none was allowed although there was much from abroad. By 1993 free market conditions meant that competition in Poland was allowed. However, none had been established due to the capital intensive nature of the industry combined with the lack of finance available within Poland. This situation looks unlikely to change so that there will be no need for POL to reposition on this front. Potential new competition due to the encouragement of free market principles in Poland might include Polish Steamship Company and shipping companies in other ex-CMEA member countries setting up services on the North Atlantic.

Promotion

1. Sponsorship and hospitality

The amount of sponsorship and hospitality carried out by POL was low for 1988, and even lower for late 1992/ early 1993. The limited amount of activity in this area - which concentrated upon dining with important customers - had to be reduced by early 1993 due to financial constraints caused by heavy company losses.

2. Advertising

Advertising is a major method of promotion. The extent of advertising carried out will be affected by the budget, and the depth of advertising reflected in the media used as well as the frequency and types of advertising carried out. POL have always advertised to promote the company and to gain customers and this looks likely to continue especially as they are increasingly working on a western basis. In 1988 the advertising budget did not exist as such; the state would have paid all expenses incurred in advertising in the general and shipping press. In 1993 this was the responsibility of the company and would have to be paid for from the profits. The shortage of funds within the company led to greater targeting and more selective advertising; for example POL withdrew from advertising in the general press and limited their chosen media to the shipping press. Advertising in 1993 was carried out only occasionally and was related to special reports and surveys. Throughout their advertising campaign POL have continually concentrated on making their presence felt within the market, and have attempted to provide information rather than use persuasion in their advertising strategy. Changes on the advertising front include the reduction in budget leading to more selective advertising which reflects a repositioning exercise concentrating attention to specific markets where short term returns are likely to be higher.

3. Image

The development of the corporate image is one of the underlying concepts of positioning. POL have always had a corporate image and this strategy is likely to continue to keep in line with western operators. The corporate image is in keeping with the styles of the other Polish shipping companies, and was established by the state when the companies were formed. However in recent years there has been the opportunity to change and although no action has been taken as yet, changes are planned, particularly following privatisation at some future date. As the company splits, the new joint stock companies created from POL are likely to want to project new individual images. However, there is a paradox here as POL needs to project a new image and dispel the old state related one for commercial reasons, but this will involve considerable expense, currently unavailable, so that much needed changes may have to be delayed, postponing commercial/ financial viability of the company.

4. Representatives

Shipping companies often use representatives as a means of projecting their company into foreign parts of the world. The number of representatives employed provides a measure of the attention given to the promotional element. This number may then be used for comparison purposes. Another measure will be the spread of location of the representatives and similar information for agents if applicable. POL have always used both representatives and agents. In early 1993 they employed 37 representatives based in the following countries:

Austria, Belgium, Canary Islands, Czech Republic, Slovakia, Denmark, Finland, France, Germany, Italy, Portugal, Spain, Sweden, UK, Russia, Egypt, Ghana, Libya, Morocco, Nigeria, Tunisia, Brazil, Peru, USA, Canada, Bangladesh, India, Indonesia, Japan, Singapore, Sri Lanka, Syria, Thailand, Vietnam, New Zealand.

At this time POL also used 247 agents spread worldwide. Although there are no specific changes planned, numbers and locations of representatives and agents are to be reviewed at a future date. The locations are likely to alter as markets do. The situation again presents a paradox as more representatives are needed for greater market penetration yet there is no money available to employ them.

5. Market research

Market research is an increasingly important method of facilitating promotion, used in numerous industries with great success. The extent of market research carried out will be indicated numerically by the market research budget and by the number of people employed in the marketing department. In 1988 POL carried out market research in collaboration with the University of Gdansk. This was entirely state funded so that the actual budget figure is not available, if any transaction of money occurred at all. In 1993 market research was still being carried out although the budget was now a company responsibility. Although no figures could be produced it was clear that the amount of funds available was highly restricted. In the future it is likely that research will be further reduced due to money shortages, although this conflicts with the planned privatisation which will require increased market research, thus presenting another paradox. The only solution in the short term has been greater targeting of market effort and more selective research priorities.

6. Public Relations (PR)

A further method of promotion is through the use of PR. This may be measured in terms of a PR budget, and for example through the variety and number of promotional items produced. In the past this was entirely state funded so that the budget remains unknown if one existed. In 1988 there were a wide range of promotional items, such as wallets, clocks, ties and scarves and these were produced in large quantities. With the introduction of free market operations the company has become responsible for the budget which consequently has been considerably reduced. The variety of promotional items has been rationalised and numbers distributed are much more limited. PR material is now only presented to VIPs. The amount produced can only fall in the foreseeable future, at a time when PR might play a valuable part in promoting company activities and image in a free market situation.

7. Directories

Directories are widely used in the shipping industry as a method of promotion. The number and spread of entries will vary between companies and from year to year, depending upon previous success rates, marketing strategies and market climates. POL have appeared in major world shipping directories such as Lloyds Shipping Directory, Fairplay, and Containerisation International, whilst daily shipping movements are shown in Lloyds Loading List. This remained unchanged between 1988 and 1993 and is unlikely to change - with no positional impact. Entry in such directories is standard practice for all international shipping companies.

8. Literature

The production of marketing literature provides a tried and tested method of promoting the company and is widely used in competitive industries. It may be produced in various forms and languages and sent to a range of destinations. The basic form of measurement is a mix of quality and quantity. The situation as regards POL's literature remained mostly unchanged between 1988 and late 1992/ early 1993. POL continued to produce literature in the form of brochures and booklets detailing the broad range of services, plus details of representatives and agents. These are all written in English - as it is the international shipping language - and they are distributed worldwide to all representatives and agents. Possible changes are that despite the pressure from privatisation to increase promotion, money shortages will lead to a reduction in the quantity of literature produced. There may also be greater targeting of the literature, new styles and the inclusion of advertisements from other companies to help cover costs.

9. Agencies

Advertising and PR agents are often used to promote the company as they have the necessary experience and expertise within the specific market. Various types of agency exist and their usage is reflected in the budget set aside for agency expenditure. POL have never used advertising or PR agencies and this looks unlikely to change especially with increasing pressure upon capital resources. There are therefore no impacts in terms of positioning.

10. Exhibitioneering

Exhibitioneering is sometimes used as a method of promotion in the maritime industries. Where this is carried out it may be quantified by counting the numbers of exhibitions attended and their locations. In 1988 POL attended numerous exhibitions on an international basis. With the company's severe financial problems in early 1993 this practice was suspended, although the suspension may be temporary with privatisation placing pressure on promotional activities to be resumed.

11. Press releases

Press releases provide a form of promotion frequently utilised by the shipping industry. Again the number, languages and destinations of these may vary and basic measurements and details provide comparable data and information. POL's policy as regards press releases remained unchanged between 1988 and early 1993. Press releases have always been widely issued and this looks set to continue. They are generated in variable numbers, and are produced in Polish and English. This is standard shipping company practice and is unlikely to change even with the advent of privatisation and exposure to the free market. There are therefore no impacts for the company's positioning strategy.

Physical evidence

1. Containers

Containers are an easily recognisable form of physical evidence for a liner company. Records of the number and types owned by the company will enable year to year comparisons, and suggest to some extent the changing position of the company. Many companies hire containers which may be painted in the company's own colours to increase physical evidence and aid promotion. Here records of numbers and types will also help to create a picture of company policy. In 1988 POL operated 32,374 TEUs, of which 14,260 were owned, the remainder leased. By 1993 this figure had almost doubled as POL operated some 60,000 containers, of which 85% were standard TEUs. Rented and leased containers made up approximately 40% of the total, the rest being owned outright. In the future POL expect to reduce the number of containers owned and to increase short term leasing in the light of more competitive and freer markets. All owned containers were painted in company colours - but none that were leased or rented.

2. Vehicles

Vehicles provide another element of physical evidence that may be simply measured by counting numbers owned by the company and noting the types and manufacturers. In 1988 POL's fleet of vehicles comprised a mixture of Eastern and Western manufactured lorries in substantial numbers, details of which remain unknown. In 1993 the company owned 120 Swedish made Volvos only, noted as a substantial reduction in the total numbers from 1988. In the future with the advent of privatisation POL may purchase western rather than eastern manufactured lorries because of their superior quality and their ability to raise the company's profile. As the company is no longer controlled by the state it is no longer required to buy East European goods. However, one of the drawbacks of privatisation and the withdrawal of the state will be the reduction in finance available for investment in new (western) goods - particularly in the short term. It is planned that the advent of privatisation will see the inland transport section of POL sold off as a separate enterprise so that POL will be free to use other trucking companies, within a competitive environment. By 1993 there were already plans to form a joint venture between POL lorries and a Danish enterprise. Company position is thus one of moving from a state driven mixed vehicle fleet of mixed quality, to one western dominated but of a higher image - but financial constraints may delay this move.

3. Company offices

Company offices are perhaps the most recognised form of physical evidence. Measurement of these involves looking at numbers, sizes and locations. In 1988 POL had one company office which was their Head Quarters in Gdynia, and they owned a number of depots and warehouses. By 1993 there were two offices, the main office at Gdynia and the office for Euroafrica at Szczecin, as well as the other warehouses and depots. In the future, with privatisation, the company is expected to split into various sections so that the number of independent companies will increase although the locations for these companies' offices remains unknown. It is likely that some locations will close and new ones will open up, with a diversification of activities at each. Although not reflecting a deliberate policy of positional change, changes elsewhere (for example through company diversification and privatisation) will have an impact upon company office location in the market.

4. Ports served

The ports served by a shipping company provide another form of physical evidence within the market place. These may be measured by counting the number served and looking at their locations in order to provide comparable year on year information. In 1993 POL served over 100 ports throughout the world. In the past the number of ports served was higher as POL was obliged to serve CMEA ports such as Havana (Cuba) which commercially are no longer profitable enough to warrant serving at all, or as frequently (eg Vietnam). The company is likely to rationalise services further as it concentrates more fully on the most profitable areas. These significant changes represent positional change through physical evidence. On the North Atlantic trade POL by 1993 called at the ports of Gdynia, Bremerhaven, Le Havre, Halifax, New York, Baltimore, Wilmington, New York, Le Havre, Rotterdam, Bremerhaven, Gdynia. They could also provide a through transport arrangement to/from any point in the USA/ Canada/ Europe and they laid on feeder vessels from Gdynia to Hamburg and Bremerhaven return, and from Le Havre to Tilbury return. This did not in itself reflect any real change from 1988 - and it is in other markets, such as South America and the US Gulf, that the company has shown the majority of port positional change.

5. Routes served regularly

The routes served regularly provide a form of physical evidence in a similar manner to ports served. They are most easily 'measured' by looking at their number and details. In 1988 POL offered services in eight categories these being, services for America, Australia, Europe, West Africa, Asia, South Pacific, Mediterranean, and Red Sea. More specifically these included operations between the Baltic/ North Europe and ECNA, ECSA, Mid East/India/Pakistan, Red Sea, N Africa/ Near East, and Mediterranean/ Near East. From North Europe services ran to and from WCSA, Australia, Mid East/ India - Far East/ Japan, and South Pacific. POL also offered some landbridge/ minibridge services as well as providing a number of services between Poland and North Europe and Scandinavia, and one service between Baltic/ N Europe and West Africa.
In 1993 POL regularly served 11 routes, these being:

North Atlantic Con-Ro Service - one sailing per week
US Gulf, South Atlantic Container Service - every 11 days

South America West Coast Line - every 3 weeks
Central America, Caribbean and US Gulf Line - every month
East coast and South Africa Line - every 8 weeks
Mediterranean Service-Mars Line - every 10 days
South America East Coast Line - every 3 weeks
Asia - JOS Service - every 7 days
South Asia - Indonesia/ Vietnam Line - 1-2 every month
South Asia - Bengal Bay Line - every month
Australia Line - every 15 days

Pre 1988 the number of routes served was higher as POL regularly served CMEA routes which have been dropped as they are not commercially sound. However, changes in ports served both in 1988 and early 1993 reflect better the route changes. This movement is likely to continue which reflects a significant and deliberate positional change.

Process

1. Conference and consortia

Conference and Consortia membership is a significant aspect of the process element of the marketing mix. In order that year on year comparisons might be made, they may be measured in terms of which conferences or consortia the company is a member of and at what level of membership. In 1988 POL was a full member of one Far East and two South American conferences, and notably not a conference member on the North Atlantic. By 1993, conference membership had been dropped generally due to market conditions. This included membership of the two South American conferences and suspension of the Far East conference. However, when the TAA was formed in 1992, POL became an unstructured member - reflecting their previous non conference membership in this market, but also their desire to move up market from the lower image of an ex-CMEA operator and their freedom from state requirements.

2. Guaranteed traffic

In 1988 a considerable amount of the services sold for POL involved the state's guaranteed traffic and that for other CMEA governments. By early 1993 guaranteed traffic was considerably reduced with less coming from the state and the CMEA market lost entirely. Although there are likely to be further reductions some guaranteed traffic will always remain (eg military equipment and state sensitive goods). POL will need to compete for other cargoes to make good the shortfall of traffic from its operations, requiring positioning through process change, more agency activity and moves towards more indirect methods of sales, plus more active marketing, better computer systems and higher quality services.

3. Collaboration

Collaboration is another process element which can be 'measured' by examining the types of collaboration taking place, the modes and the extent of their operations. POL has long owned a trucking company which comprised 120 Volvos in 1993. Future plans include privatising this company so that POL will no longer have any direct road collaboration. Collaboration with rail has continued through from 1988 and looks set to continue, although there is likely to be a change of emphasis as the transport market as a whole becomes freer and the need for a more flexible service emerges - particularly related to JIT issues. In the past all elements of transport were state owned and there was forced collaboration. By 1993 companies/modes had some independence, but collaboration was continuing to some extent due to continued practices of tradition. In the future, collaboration is likely to continue, but this will be due to market forces so there will be variations emerging. These changes affect the process element of the marketing mix and will force POL to reposition to some extent. For example there may be better collaboration with private road industry and less domination of rail - reflecting the moves towards a more industrially responsive transport network.

Place

1. Origins and destinations served

Origins and destinations served are an element of place that may be measured by number, but they will also vary in frequency, location, and in other factors. In 1988 POL called at approximately 100 ports, operating 17 different liner services. In late 1992/ early 1993 POL regularly called at a similar number of ports, although policy dictated that their exact number and locations were related to demand. They operated 21 liner services including 14 trans ocean lines. Although they operated on a worldwide basis, their main trades were the North Atlantic and the Far East. In the past, POL services would have been controlled by the state. In 1988 POL still operated on a worldwide basis but they also served CMEA and associated countries' ports. In future further rationalisation is likely to decrease the number of places served and service frequency will alter according to market demands. These changes in the place element both require and reflect significant repositioning by POL.

144

2. Computer reservation systems

The presence of computer reservation systems will provide another indication of positional change in place. In the past POL had no computer reservation systems but by 1993 they were used by the majority of representatives and agents. This represents an alteration to the method of selling space on board vessels and makes access to the market at home and abroad easier. It also enables POL to provide a better quality service for example through faster, more accurate bookings. This in turn has enabled POL to alter significantly its place position in the market, with more flexibility in sales location and more accuracy and speed in confirmation.

People

1. Qualifications and training

Qualifications provide a useful measure of the quality of personnel employed by a company and an indication as to the company's position. They can be compared by looking at the level of qualified staff within the company, the sources and types of qualification and the importance attached to them. In the past the only requirement for POL management was seafaring experience and managers would usually have been ex-seafarers. For seafarers employed by POL, Polish government controlled qualifications have been required throughout the company's history. In 1992 seafarers still qualified at nautical school in Gdynia but POL also had begun to provide in house training such as management courses with consultancy companies, and some out house training courses. There has also been some progress towards foreign training using money from sources such as the European Community, and many employees have attended short courses on Information Technology at the University of Gdansk. In future employees are just as likely to be non seafarers who are management business trained to University level and non shipping specialised. The new training techniques will bring new perspectives into the company and present a company with an image of higher quality. POL has already begun to reposition on the qualification and training aspects of the people element to a radical extent in comparison with the approach before 1988.

2. Employees

Employee numbers present the most basic picture of the people element of the marketing mix and the company's position. In 1992 POL employed 5,800 staff, 4,500 of these worked at sea and 1,300 ashore. In the past these figures were notably higher, as approximately 8,000 people were employed by POL in 1988 with around 6,500 at sea. Figures have fallen due to the creation of a freer market which has emphasised over employment within the company and necessitated rationalisation. However, when full privatisation takes place and the company is allowed to make its own decisions about dismissals, then it is expected that approximately 2,000 further redundancies will take place. In 1992 POL was still forced by the state to employ staff for which it had little need and could ill afford to pay. Future rationalisation, albeit forced upon them from external forces, will leave POL as a leaner and fitter company that can occupy a more competitive position in the market place.

3. Management structures

Management structures within a company are an integral aspect of the people element and have significant impacts upon the operation and success of a company. Their details can be noted to provide year on year comparisons and to indicate change within the company. In 1988 POL was state managed through an overall state appointed board. By 1992 the control was held by a company board, and the state POL department with specific responsibility for shipping operations was abolished. Although still state owned the company had much greater independence. In future, disaggregation is expected as the company is gradually privatised and split into a number of profit centres. Generally there is an overall move away from one of administrative management towards capital management. In early 1993 a completely new management structure was established giving even greater company and departmental independence in terms of budgets and decision making. This created major changes in attitudes and approach which directly affect decision making and the ability of staff to react to changing market situations.

4. Unions/ Workers councils

Unions often play a very significant part in the people element and are best measured by examining their membership levels and their role within the organisation. In 1988 the majority of POL workers belonged to a union. These were mainly Solidarity and OPZZ, with some workers belonging to the Polish Officers and Crew Members Unions. The unions were seen as having an important role within the company decision making process, to the extent that they formed a highly significant element of the management process. By 1992 membership levels had fallen to approximately 50% and unions were becoming less popular due to their association with the old Communist regime. In future membership is expected to fall further, and the unions are likely to decline in importance as market forces take over. This will leave greater power with the company management and will enable them to react more quickly to new situations. Workers Councils - representing the employees within the company, remain a significant part of the decision making process in state owned companies such as POL - where decisions on pay, redundancies and working conditions all need their approval. Privatised companies have no Workers Councils and thus are freer to react to market needs.

Product

1. Speed

The product of a liner shipping company is the transport of goods from origin to destination. One way in which this may be quantified is to measure the speed of the service. In 1992 POL's North Atlantic operations included services from Gdynia to New York in 10 days at a maximum speed of 21.65 knots. In 1988, despite similar vessel capabilities - 21.7kts - the service was notably slower, crossing the Atlantic on average in 12 days, and it was a more variable service reflecting the lower price image. In the future the speed of service will vary in direct relation to areas such as demand and fuel costs, as the company is further exposed to market forces.

2. Safety

The examination of the safety records for a shipping company provide another measurement of the product. No data is available concerning past safety records and container losses. In 1992 POL maintained that they had adopted all international regulations and that they had good safety and casualty records, a view supported by Lloyds List (15/01/93). This record will have to be maintained if not improved upon in order to attract new customers and expand market share. However, due to limitations upon information, few conclusions can be reached.

3. Reliability

Reliability is an important element of liner shipping's product. It may be measured in terms of number of late deliveries within a given space of time. No data is available on the reliability of POL's services due to the commercial sensitivity of such information. However, now that POL is operating in a truly free market reliability of service will be of greater importance to the company and will take on a greater significance in the company's external image and therefore position.

4. Frequency

The frequency element of product is most simply measured by looking at the time between services. In the case of this research the service is between North Europe and the US east coast. In 1992 POL ran one service per week on the North

147

Atlantic trade - which had not varied since 1988. Thus no positional change.

5. Quantity

The quantity aspect examines the amount of cargo capacity provided by POL on the North Atlantic trade. To enable comparisons to be made a standard measure of TEU's is used. In late 1992 POL was using 4 sister vessels of 1,424 TEU on the North Atlantic trade, which represents little difference over the 1988 position when four vessels of 1,417 TEU were used.

6. Ship Types

Ship types are another element of the product. They can be divided into basic categories, and counted to provide numbers of each type as well as a total fleet number and dead weight tonnage (dwt). In 1988 POL owned and operated 107 vessels with a total dwt of 1,199,589. As shown earlier in table 1 these consisted mainly of DC's (40), DN's (37) and WA's (15). By early 1993 POL owned and operated just 23 vessels with a total deadweight tonnage of 273,884. These can be broken down into the following categories:

Table 12
Polish Ocean Lines fleet 1993

Type	1988	1993
DC	40	11
DN	37	2
WA	15	6
PM	2	2
CL	3	1
WL	2	1
Others	8	0

Source: Fairplay World Shipping Directories
Key to vessel types - see appendix 1

Polish Ocean Lines' statistics show a dramatic decline in the size of the fleet from 107 vessels in 1988 to 23 vessels in early 1993. This is partly due to the initial privatisation processes taking place which have split the company into a number of sectors leaving the core of POL both limited and specialised. The changes are aimed at greater efficiency.

The main changes planned for the future are an increase in container vessels, especially in larger capacity types, namely 3,000 TEU. The general cargo vessels are gradually being replaced by container vessels. These changes are a reaction to changes in the market demand, but are fairly long term and will not have a positional impact for some time. Coupled with the changes forced upon POL by the privatisation process, we can see a notable positional change reflecting a need and desire to improve quality and efficiency.

7. Ship ages

Studying ship ages helps to give some indication of the quality of the product. In 1988 POL's vessels had an average age of 13.46 years, and a number of vessels were over 20. By early 1993 the average age of vessels was down to 8.65 years due to restructuring processes taking place as noted in (6) above; reflecting the same moves towards quality and efficiency. In future the average age is likely to increase as financing problems prevent sufficient renewals and new buildings. Ageing vessels will lower the quality of the service and POL may have to reposition further in the market in order to compete on a different basis.

8. Flag

The flag under which the vessels sail is a further element of product in shipping. In 1988 all of POL's vessels sailed under the Polish flag, as the state provided no option. By late 1992 POL had flagged out 6 vessels to Cyprus. In future this practice is planned to increase to satisfy the requirements of financiers such as western banks. Flagging out has impacts in terms of quality of service, and may affect other aspects such as safety.

9. Cross trades

Cross trades are an important part of liner shipping operations, especially in East Europe where hard currency remains and has always been a significant issue in company positioning. For comparison purposes they may be measured as a percentage of the company's total activities, but the details of the trades are also relevant and should form part of the records. In 1992, 45% of POL's container activity occurred on the cross trades. In 1988 cross trades made up a slightly lower percentage and the routes would have varied as CMEA members would have been served. In future the volume of cross trade activity is expected to rise to compensate for a decrease in domestic trade as an increasing amount of goods are carried by road and rail through Germany, and as the recession continues to bite.

10. Computer tracking for containers

Computer tracking systems are another element of product. It is not readily measured but the details may be noted for comparison purposes. In early 1993 POL operated a limited computer tracking system for its containers which enabled it to provide customer information on the approximate latest known location. This was not used in 1988, but in the future it is likely that its use will be expanded to provide a better quality of service.

11. Trading terms

This is an important element of product for East European liner shipping companies. In the past Poland and other CMEA countries sent all exports CIF (Carriage, insurance and freight) and received all imports FOB (Free on board). This meant that Poland controlled all transport of goods entering and leaving the country. This had a major impact for companies such as POL who were consequently guaranteed cargoes. Since 1989 the state has given up its enforcement of this, and operations are run on a purely commercial basis. The impacts of this require POL to reposition within the market place as they will have to compete for cargoes on a western style basis.

From the framework a number of main themes stand out which have produced noticeable positional changes in POL on the North Atlantic market over the period between 1988 and 1993. Firstly there have been changes in the general shipping market which have seen many companies such as POL drop the majority of their conference memberships. There has also been a change in the profitability of routes and the greatest gains in 1993 were to be made from the cross trades - not just in terms of hard currency, but also in broad commercial terms.

Secondly, POL became an associate member of the TAA. This prevented the company from offering rebates and discounts to customers on the North Atlantic trade as it had done in the past. However, it did enable POL to raise the cost of its North Atlantic service to a fixed price set by the TAA.

Perhaps the most significant changes in POL's liner shipping operations have been produced by the events occurring in Poland in recent years. These events have caused fundamental changes at every level. The collapse of the CMEA means that POL services to ex-CMEA members are no longer part of a barter system but are paid for by bank transfer in convertible currency. It also means that POL has significantly less guaranteed traffic, which is added to by the fact the trade terms are now negotiable and Poland will not be automatically responsible for the transport of all its imports and exports. In early 1993 POL no longer segmented the market merely on a basis of CMEA or non CMEA membership but had begun to work on a more western style basis. The freer conditions within Poland have also presented a long term threat of Polish competition to POL.

150

Significantly the changes have also meant that the role of the unions has become diminished and membership has fallen noticeably. The partial convertibility of the Polish currency has changed the emphasis of POL operations; it no longer needs to concentrate as much on earning hard currency but can focus on serving the most profitable routes.

The diminishing state control over POL has caused significant operational changes. By 1993 the company was free to chose the ports and routes it served, it was free to flag out, it was able to purchase western products such as vehicles and computer equipment, and it had responsibility for prices, income, and staff including qualifications and training.

Reducing state control has also had significant impacts for management and training issues, calling for changes at every level, from attitudes to decision making processes.

The decrease in state control was accompanied by the removal of most subsidy to the company. This, combined with the decline in guaranteed traffic meant that POL needed to increase its profits significantly just to survive. However, after making substantial losses in 1991 and 1992, the company had no funds. Although new vessels were urgently needed along with improvements to quality of service, the creation of a new image, increased advertising, representatives, agents, public relations, and exhibitioneering, POL was in fact having to rationalise these aspects in order to cut costs.

It is planned that POL will be fully privatised eventually. This is to be achieved by splitting the company into a number of profit making centres. This will have impacts upon the company's management structures, and the number and location of offices. One of the main initial impacts of privatisation will be the redundancies of large numbers of staff. Dismissals will only be allowable after privatisation takes place and POL estimate that 2,000 may be necessary by that time.

In the near future, although POL may have liked to reposition in the market place, a lack of funds will prevent them from doing so in the way needed. This presents a paradox as without adequate repositioning POL's situation is unlikely to improve sufficiently for the much needed extra profits to be generated.

To facilitate analysis POL's 1992 position can be outlined further using a SWOT analysis to provide structure, derived from earlier discussion.

Strengths

POL's market strengths may be summarised as follows:
- cheap labour relative to western operators, offering cost advantages
- skilled and experienced labour, especially seafarers who gain Polish government controlled qualifications from the nautical school in Gdynia
- a strong Polish maritime tradition
- an existing market niche and existing cross trade market positions

- advantages of geographical location, which include their proximity to Germany, and their positions between EC and East Europe, and between Scandinavia, South East Europe and the Middle East
- previous experience of surviving in a western world market and of earning hard currency
- general political/social reputation in the west compared to other East European operators
- membership of the TAA
- good relations with the European Community, compared to other East European countries

Weaknesses

The weaknesses of POL's operations may be summarised as:
- poor general image amongst the shipping fraternity, of cheapness, low quality, and low level technology
- lack of finance available to improve upon image combined with problems of financing/obtaining loans
- continued shortage of hard currency despite some improvements, affecting operational decisions
- poor political prospects, including instability of government and potential reversal to old regimes
- banking and accounting inadequacies within Poland
- lack of modern commercial and management skills amongst employees, as well as continued entrenched attitudes
- state withdrawal and industrial restructuring leading to a loss of government coordination (eg. intermodalism) without a market replacement
- peripheral infrastructure and communication problems
- domestic financial weakness and high taxation

Opportunities

POL's opportunities can be summarised as:
- joint ventures; western interest and involvement is relatively high, offering potential benefits such as improving management skills and technical efficiency
- privatisation could also lead to diversification, and the opening up of new ideas and capital
- cheapness needs to be exploited in a way that does not detract from image
- future European Community membership and physical integration into Europe
- western aid (EC, IMF, WB) and relaxation of debt requirements
- increased East West trade as the CIS develops

152

- proximity of the Scandinavian countries, especially as they move towards the European Community
- Polish economic growth
- flagging out now permitted, affecting a range of issues including investment opportunities, costs and image
- size and quality of unemployed labour force

Threats

The threats to POL can be summarised as:
- bankruptcy or western takeovers
- political and social change, reversal and instability
- inflation and failure of the Polish economy
- new competition - both Polish and foreign - from other companies and other modes
- instability in the ex-USSR
- Single European Market becoming a fortress
- costs of increasing safety and environmental standards

Despite the daunting combination of these threats with the numerous weaknesses - for example a lack of finance - POL does have a number of strengths and significant experience which may enable it to take full benefit of opportunities such as privatisation and joint ventures.

Having established the positions for POL in 1988 and 1992 we will go on, in chapter seven, to make further use of the framework technique to establish the positions of some of Polish Ocean Line's competitors on the North Atlantic trade, for comparative purposes.

7 European Community shipping

In this chapter we will examine the positions of Polish Ocean Lines' competitors on the North Atlantic west bound trade. This will be carried out for the period of end 1992/ beginning 1993 as it is only necessary to assess the Community company positions at the end of the East European transitional phase. The assumption is made in this positional comparison, that whilst POL's position is likely to have altered from 1988, due to external pressure as discussed earlier, and caused by social, economic and political change in East Europe, that of the European Community competitors will have remained relatively constant. Hence positional comparisons are only needed for 1992/1993 between the competitors based in the European Community and POL. Clearly each of these companies will be affected by developments in European Community shipping policies, and a summary of these, together with relationships with East Europe is provided in Appendix 3. The Community companies operating on the trade have been mentioned previously, but it is necessary now to examine the companies in rather greater detail, and in the context of all North Atlantic operations, to establish the market position they have taken up and to place POL's activities in the North Atlantic market into context. Positioning is clearly a relative concept and means little without assessment of competitors' positions.

We will now look at a brief summary of companies from all countries involved in the westbound trade at the end of 1992 (a key to shipping abbreviations used in the following section can be found in appendix 1):

Atlantic Cargo Services AB

This Swedish company was owned by Benfo BV (Rotterdam). The line utilised 4 multipurpose vessels able to load containers, breakbulk, and heavy lift/project cargo. It operated a Scandinavia/North Europe-USEC/GC service every 12 days using 4 chartered in vessels with the following specifications:

Number	Type	Speed	TEU capacity
2	BC	15	1,370
2	BC	15	1,434

Ports of call on this service were Bremen, Rotterdam, Le Havre, Charleston, Port Everglades, Houston, Mobile. The company operated 10,450 TEUs, 4,850 are owned and 5,600 leased. Overall, Atlantic Cargo Services was a relatively small player in this market.

Atlantic Container Line

Based in the United States, this company was a wholly owned subsidiary of the Transatlantic Shipping Company Ltd. On the North Atlantic trade it operated a weekly service calling at the following ports:
   Antwerp, Bremerhaven, Rotterdam, Le Havre, Halifax, New York, Norfolk, Baltimore, Philadelphia, New York, Halifax, Antwerp, using the vessels of Hapag Lloyd as part of a joint service.
The fleet statistics were as follow:
   7 vessels in total (4 chartered in), of which 5 operated on the ECNA/Europe/Scandinavia service which are all Ro-Ro with cellular space, speeds of 18kts and capacities to carry between 2,911 and 2,915 TEU.
The company operated 50,778 TEUs in total of which 92% were owned and the remainder leased.

CGM

In 1992 the French CGMF holding Company controlled the groups' four major subsidiaries which included 99% of Compagnie Maritime Generale (CGM). CGM was involved in container services independently and through various space charter agreements. The company formerly operated two North Atlantic services through a slot charter agreement with the ACL consortium in coordinated sailing agreement with Hapag Lloyd, before it launched its own operation. CGM were forced to withdraw from the four North Atlantic routes it operated before the end of 1992 due to continuing losses (Containerisation International 1992). As a result CGM have been eliminated from the study, but clearly its previous operations need to be recognised to assess market condition.

Cho Yang

Registered in South Korea, Cho Yang worked in conjunction with DSR-Senator, to run the following services which covered the North Atlantic:

N Europe/UK/Far East/Japan/USWC/Panama/USEC (weekly eastbound) calling at Bremerhaven, Hamburg, Rotterdam, Antwerp, Felixstowe, Singapore, Hong Kong, Kaohsiung, Busan, Osaka, Yokohama, Oakland, Long Beach, Cristobal, Jacksonville, Wilmington, Philadelphia, Bremerhaven.

N Europe/UK/USEC/Panama/USWC/Japan/ Far East (weekly westbound) calling at Bremerhaven, Hamburg, Rotterdam, Antwerp, Felixstowe, Le Havre, Philadelphia, Wilmington, Jacksonville, Cristobal, Long Beach, Oakland, Yokohama, Kobe, Busan, Keelung, Hong Kong, Singapore, Bremerhaven.

The company operated 14 vessels of which 1 was chartered in and 6 operated on the Europe/Far East/USWC/EC/Europe. The latter were all fully cellular vessels with the following speeds and capacities:

Number	Speed (kts)	TEU capacity
1	21	2,650
1	20.5	2,288
4	22	2,698

Containers operated numbered 22,870 TEUs of which 17,369 were owned and 5,501 leased.

Contship UK Ltd

UK based Contship operated as part of its services in 1992, Ocean Star Container Line. Ocean Star operated a round the world service employing five BV 1000 self sustained container vessels of 1,000 TEU capacity and five ro-ro vessels of 1,600 TEU capacity with a nine day frequency. Ports of call included La Spezia, Genoa, Marseilles, Felixstowe, Hamburg, Rotterdam, Antwerp, Dunkirk, Le Havre, New York, Norfolk, Savannah, Papeete, Auckland, Sydney, Melbourne, Brisbane, Noumea, Lae, Jakarta, Singapore, Port Kelang, Colombo, Alexandria. Containers in service numbered 4,800 20 and 40 ft dry cargo containers.

Deppe Line

The German registered Deppe Line operated a North Europe-USEC/GC service every 6 days, which called at the ports of Rotterdam, Bremerhaven, Felixstowe, Bremerhaven, Hampton Roads, Charleston, Miami, Galveston, New Orleans, Hampton Roads. This was a joint service with Lykes and used one chartered in

SC vessel with a speed of 17.5kts and capacity of 1,095. Containers operated numbered 3,775 TEU.

DSR

In 1992 this German based company operated a wide range of services and belonged to a consortium, with Senator and Cho Yang, which operated an east/west bound around the world service with ports of call as listed under Cho Yang. DSR operated a total of 55 vessels with ranging capabilities and reported a total of 20,000 container units in service.

Evergreen

Evergreen is a Taiwanese company which operated a fixed day weekly service round-the-world eastbound and a round-the-world westbound every 6 days. Ports of call on these services were as follow:-

RW westbound - Tokyo, Nagoya, Osaka, Busan, Keelung, Kaosiung, Hong Kong, Singapore, Colombo, Hamburg, Thamesport, Rotterdam, Antwerp, Le Havre, New York, Norfolk, Charleston, Kingston, Panama, Los Angeles, Tokyo.

RW eastbound - Singapore, Hong Kong, Kaohsiung, Keelung, Busan, Hakata, Osaka, Nagoya, Shimizu, Tokyo, Los Angeles, Charleston, Baltimore, New York, Le Havre, Antwerp, Rotterdam, Thamesport, Hamburg, Colombo, Port Kelang, Singapore.

The company owned 51 vessels of which 3 were chartered to Uniglory and 24 operated on round-the-world services; the latter were all fully cellular vessels with speeds of 20.5 and 20.7kts, and capacities of 3,428 (11 vessels) and 2,728 (13 vessels). Containers operated numbered 215,075 TEU.

Hapag Lloyd AG

This German company was involved in a number of joint space charter services and was a partner in the Tri Continent service operated in conjunction with NYK and NOL. On the North Atlantic trade Hapag Lloyd operated a weekly service calling at the following ports:

Antwerp, Bremerhaven, Gothenburg, Rotterdam, Le Havre, Liverpool, Halifax, New York, Baltimore, Portsmouth, Montreal and Toronto - Hamilton and Quebec are served overland. Feeder services are provided to and from Dublin, Boston (Mass) and Portsmouth NH. This was a joint service with ACL.

The fleet statistics were as follow:

28 vessels (6 chartered in, 3 chartered out, 1 owned by Hapag Lloyd International Singapore chartered out)
3 operated on the Europe - ECNA which were all fully cellular with speeds of 20.5kts and capacities of 2,594 TEU.
The company operated 110,127 TEU.

Independent Containerline Ltd

The North Europe-USEC was the only service operated by this American based company. Ports of call were Antwerp, Chester PA, and Richmond VA. It used three chartered in FC vessels and ran every 9 days. The vessels all operated at 16kts, 2 holding 918 TEUs and the other holding 754. Containers operated numbered 5,470 TEU.

Lykes Lines

This American company was a subsidiary of the Interocean Steamship Corporation. It operated a USGC/EC - North Europe/UK service every 6 days calling at Houston, Galveston, New Orleans, Baltimore, Norfolk, Antwerp, Bremerhaven, Felixstowe, Le Havre, Norfolk, Galveston. This was operated through a joint/vessel sharing agreement with Deppe. Of the 34 vessels operated, 10 were chartered in, 4 chartered out, 5 operated on the service which covers the North Atlantic, and 3 operated on various services as required. Details of the latter 8 vessels were as follow:

Number	Type	Speed (kts)	TEUs
3	FC	18.5	1,093
2	FC	19	2,411

Variable Operations:

Number	Type	Speed	TEUs
2	SC	17.5	204
1	BB	20	61

The company operated 14,888 TEUs of which 14,848 were owned and 40 leased.

Maersk

The Danish company Maersk, belonged to the AP Moller Group and operated a fleet of fully containerised and ro/ro vessels with cellular space in a global network of liner and feeder services. It operated a N.Europe/UK - USEC/WC fixed day weekly service calling at the following ports:

Le Havre, Felixstowe, Rotterdam, Bremerhaven, Halifax, New York, Norfolk, Charleston, Long Beach, Oakland, Long Beach, Miami, Charleston, Baltimore, New York, Le Havre with other American and European ports served via feeder/inland transport services. The service was fully integrated with the FE/Japan/USWC/EC link.

The fleet statistics were as follow:

55 vessels (16 chartered in, 2 long term leased from DMK leasing, 5 bareboat chartered from DMK leasing, 5 owned by Maersk Singapore Ltd).

12 vessels were operated on the North Atlantic service, these were all fully cellular 10 of them having speeds of 24kts and capacities of 4,000 TEU, the remaining two having speeds of 23kts and capacities of 3,000 TEU.

Maersk operated over 125,000 TEU containers.

Mediterranean Shipping Company SA

Registered in Switzerland, the company operated a fixed day weekly service on the North Europe/UK - USEC route which called at the following ports:

Antwerp, Hamburg, Bremen, Felixstowe, Le Havre, Boston, New York, Baltimore, Newport News, Antwerp. Joon Liner services Express Line has a slot charter arrangement with the service.

The company controlled 35 vessels of which 2 were chartered in and one was chartered out. Three vessels ran on the North Atlantic service, 1 was fully cellular and could operate at 20kts and hold 1,300 containers, and 2 were ro-ro with cellular space with speeds of 18.6kts and 1,146 capacity. Containers operated numbered 55,000 TEU.

Nedlloyd Lines

Based in the Netherlands, the company was an operating division of the Royal Nedlloyd Group NV which also owned KNSM and had a 50% equity interest in North Sea Ferries. On the North Atlantic it operated a fixed day weekly service calling at the following ports:

Bremerhaven, Felixstowe, Rotterdam, Le Havre, Boston, Port Elizabeth (cargo for New York), Norfolk, Bremerhaven. Operates in joint schedule and space charter arrangement with P&O containers and SeaLand.

Fleet statistics were as follow:

59 vessels (3 co-owned with P&O and Swire, 8 chartered in, 2 owned and operated by SeaLand as Nedlloyd contribution to the North Atlantic services, 2 jointly owned with CGM, 1 co-owned with P&O containers (Pacific), 1

chartered out, 2 operated by Sinbad Lines (Hong Kong)).
On the North Atlantic the company operated 2 fully cellular vessels with speeds of 19kts and capacities of 3,546 TEU.
The company operated 145,000 TEU in a combination of owned and leased units.

Neptune Orient Lines (NOL)

Neptune Orient Lines had its head office in Singapore and associate offices in Australia, Indonesia and the UK. The fleet list was as follows:

CL	12	TN	2
WL	1	TR	6
CN	5	BN	2
CF	2	BS	7
DN	6	TO	11

The North Atlantic route was operated by NOL in conjunction with NYK and Hapag Lloyd. Ports of call were Antwerp, Bremerhaven, Rotterdam Thamesport - New York, Savannah, Los Angeles, Oakland - Tokyo/ Yokohama, Kobe - Hong Kong - Kaohsiung. Containers operated numbered 16,599.

Nippon Yusen Kaisha (NYK)

NYK had its head office in Tokyo, Japan. It operated 327 vessels of which 62 were owned. Containers operated numbered 150,451 TEU, comprising 97,632 owned and 52,819 leased. NYK was a partner in the weekly Tri Continent service operated with NOL and Hapag Lloyd. Ports of call were Antwerp, Bremerhaven, Rotterdam, Thamesport - New York, Savannah, Los Angeles, Oakland - Tokyo/ Yokohama, Kobe - Hong Kong - Kaohsiung.

Orient Overseas Container Line

Based in Hong Kong, this company was a part of Orient Overseas International Line Limited. It operated a fixed day weekly service between UK/North Europe - ECNA calling at the following ports:-
Halifax, New York, Norfolk, Charleston, Felixstowe, Zeebruge, Bremerhaven, Felixstowe, Le Havre, Halifax.
The fleet consisted of 33 vessels of which
1 is co-owned with Kline and CMB; jointly chartered with Neptune Orient Lines (NOL); operated by NOL

160

4 are chartered out
1 is chartered in
1 is co-owned with Canada Maritime
4 fully cellular vessels operated on the UK/North Europe - ECNA, these had the following speeds and capacities:

Speed (kts)	TEU
21	2,118
18.5	2,250
21.5	2,116
18.5	2,266

P&O Containers Ltd

P&O Containers Ltd was a British based company operating a N.Europe/UK - USEC weekly service calling at the following ports
Bremerhaven, Felixstowe, Rotterdam, Le Havre, Boston, Port Elizabeth, Norfolk, Bremerhaven. The service operated a joint sailing and slot exchange agreement with Nedlloyd and SeaLand.
Fleet statistics were as follow:

34 vessels	4 chartered in
	2 chartered out
	2 jointly chartered with Ben Line
	1 part ownership through share in ASCL (18.5%)
	2 part ownership through share in AAE (37%)
	3 operated by Sealand, deployed by P & O Containers in co-ordinated North Atlantic service
	3 majority ownership (74%)
	1 for AJCL service

On the North Atlantic service the company operated 4 fully cellular vessels with speeds of 19kts and TEU capacities of 3,456.
Containers operated numbered 127,000 of which 84,000 were owned and 43,000 were leased.

Polish Ocean Lines

Already discussed in some detail, Polish Ocean Lines was a state owned company which operated a weekly service between the Baltic/North Europe and ECNA calling at Gdynia, Bremerhaven, Le Havre, Halifax, New York, Baltimore, Wilmington NC, New York, Le Havre, Rotterdam, Bremerhaven, Gdynia. The service utilised four sister vessels of con-ro type, each with a maximum speed of

161

21.65kts and capacities of 1,424 TEU (320 TEU in ro/ro spaces). The total fleet consisted of 64 vessels which included a mixture of reefers, containers, con-ro, general cargo, ro/ro and feeder vessels.

Sea Land Service Inc.

The American company, SeaLand, owned 47 vessels and operated services worldwide. On the North Atlantic run it called at the ports of Boston, Port Elizabeth, Norfolk, Bremerhaven, Felixstowe, and Rotterdam. This was a weekly service using four vessels. Sea Land`s fleet consisted of 12 Atlantic class vessels (18kts, capacity 1,700 40ft units), 12 D-9J class vessels (22kts 1,236 container capacity), 4 C8 vessels (21.8kt 1,025 40ft containers), and 19 other vessels.

The company operated 110,754 containers of various types.

Senator

The German company, Senator, established an east/westbound round-the-world consortium with DSR and Cho Yang under the name Tricon, inaugural sailings commenced during January 1991.

Senator ran between North Europe/UK/Far East/Japan/USWC/ Panama/USEC weekly eastbound and between North Europe/ UK/ USEC/ Panama/ USWC/ Japan/ Far East weekly westbound with ports of call as listed under Cho Yang. 14 vessels were chartered in of which the following 13 fully cellular vessels operated on the service which includes the North Atlantic :

Number	Speed (kts)	Container Capacity
4	18	1,754
4	18	2,000
2	18	1,923
2	18	1,743
1	19	1,061

The company operated 40,735 TEU of which 2100 were owned and 38,635 were leased.

(Company information taken from Lloyds Maritime Directory 1992, Fairplay 1992-93 and Containerisation International Yearbook 1991).

There were a number of joint/vessel sharing agreements in operation on the North Atlantic in late 1992, namely:
* ACL/Hapag Lloyd operated slot sharing services in the North American trade.

162

ACL's ro-ro capacity was not included in the agreement and the two lines retained separate marketing responsibilities.

* SeaLand/P&O/Nedlloyd/OOCL - Maersk were involved in talks which centred on joining this joint service, the lines already chartered space on Maersk's sailings. The 1992 annual report of SeaLand's parent company, claimed that 'through vessel sharing and asset pooling arrangements with other container shipping companies - SeaLand has begun to reduce its cost base, whilst offering customers better sailing frequencies, port coverage and transit times'.
* Lykes/Deppe (noted earlier)
* DSR/Senator/Cho Yang (noted earlier)
* ACL/Mediterranean Shipping Company - ACL claimed that this link with MSC resulted in better utilisation of equipment with attendant decrease in costs (ACL/MSC 1992). The joint service was significant as ACL was a former 'conference' member, whilst MSC was an 'independent' carrier prior to the establishment of the Trans Atlantic Agreement.
* Singapore based NOL and Japanese NYK also set up new services on the North Atlantic in conjunction with Hapag Lloyd.

As discussed earlier this research will compare POL's operations with those of competitors on the North Atlantic trade who are based in the European Community - these being P&O Containers Ltd (UK), Maersk (Denmark), Hapag Lloyd (Germany), DSR (Germany), Senator (Germany), Contship UK Ltd (UK), and Nedlloyd (Netherlands) - and a more detailed study of these companies' operations will follow later.

The comparison is to be carried out by first examining the full scope of each company's operations within a framework identical to that of Polish Ocean Lines' in chapter six. The companies included in the framework are those noted above, ie P&O, Maersk, Hapag Lloyd, DSR-Senator, Contship, and Nedlloyd. This includes every European Community operator involved on the west bound trade within the TAA and represents both structured and unstructured members of the TAA as well as independent operators. Companies which are excluded are therefore MSC (Switzerland), Evergreen (Taiwan), Independent Container Line (USA), SeaLand (USA), OOCL (Hong Kong), Cho Yang (S.Korea), Lykes (USA), ACL (USA), NOL (Singapore), NYK (Japan) and Atlantic Cargo Services (Sweden). These are registered in countries other than those of the European Community in 1992/3. Although Deppe are a German registered company, their activities have been excluded as they offer only a limited service through a joint agreement with Lykes. Also excluded are transhipment services where the North Atlantic section forms a small part of a wider service, making the conditions of these markets irrelevant to this research.

Table 13 shows more clearly a summary of the criteria which led to company selection or rejection for the competitors' framework.

163

Table 13
Competitors on the North Atlantic 1992 - summary of country of register and level of TAA membership.

Company	Country	TAA membership	Select - yes or no
Atlantic Cargo Services	Sweden	Independent	No
ACL	USA	Structured	No
Cho Yang	South Korea	Unstructured	No
Contship	UK	Independent	Yes
Deppe	German	Independent	No
DSR	German	Unstructured	Yes
Evergreen	Taiwan	Independent	No
Hapag Lloyd	German	Structured	Yes
Independent Container Line	USA	Independent	No
Lykes	USA	Independent	No
Maersk	Denmark	Structured	Yes
MSC	Switzerland	Unstructured	No
Nedlloyd	Netherlands	Structured	Yes
NOL	Singapore	Structured	No
NYK	Japan	Structured	No
OOCL	Hong Kong	Structured	No
P&O Containers	UK	Structured	Yes
Sea Land	USA	Structured	No
Senator	German	Unstructured	Yes

Figure 5 shows the European Community framework for the selected companies which indicates the position of each chosen competitor at the end of 1992. Information for the framework was obtained during a series of structured interviews, as discussed in chapter three. These were held with senior management from the companies involved, and again an example of the interview schedule can be seen in appendix 2. The framework will now be examined in some detail, extracting the principle trends by taking each element of each 'P' in turn. It should be noted that for eventual comparison purposes the EC based competitors will generally be treated as one set rather than individually. This is possible due to their many similarities, as the majority of their operations are regulated by the same bodies, ie the TAA and European Community.

Price

1. Rebates

None of the operators studied offer rebates to regular shippers on the North Atlantic trade. Members of the TAA are not permitted to offer rebates, whilst the independent operator Contship is prevented from offering rebates by the Federal Maritime Commission (FMC) in the United States under anti trust laws. Hence for this element of price, the position for each European Community based company is identical in that no rebates are offered due to outside controls.

2. Discounts

Members of the TAA are only permitted to offer discounts in exceptional circumstances. This can occur where a shipper solely using a TAA member line such as P&O, Maersk or Nedlloyd, wishes to move to an outsider such as Contship. Where this is the case, 'independent action' is permissable with the sanction at company director level and then the express permission of the TAA secretariat. As independent operators, Contship are able to offer discounts on a commercial basis, and do so according to customer.

3. Price for transporting one TEU

The price for transporting one TEU westbound across the North Atlantic is fixed by the TAA, and this price is inclusive of all surcharges, port costs, and terminal charges. For structured members of the TAA the price level in early 1993 was US $1085. Unstructured members were set a price of approximately US $50 less. Independent operators such as Contship consequently tended to offer a price slightly below the unstructured level to attract customers.

P	Elements	Measurement Method	Contship	DRS Senator	Hapag Lloyd	Maersk	Nedlloyd	P&O
P	Rebates	Yes/No	No	No	No	No	No	No
R		Deferred Yes/No	No	No	No	No	No	No
I		Percentage	N/A	N/A	N/A	N/A	N/A	N/A
C		Other details	Rebates prevented by FMC	TAA prohibits rebates	TAA prohibits rebates	TAA prohibits rebates	TAA prohibits rebates	TAA prohibits rebates
E	Discounts	Yes/No	Yes	Only in exceptional cases	Only in exceptional cases	Only in exceptional cases	Only in exceptional cases	Only in exceptional cases
		Details	According to customer	Discounts only where shipper solely using DSR Senator wishes to move to an outsider.	Discounts only where shipper solely using Hapag Lloyd wishes to move to an outsider.	Discounts only where shipper solely using Maersk wishes to move to an outsider.	Discounts only where shipper solely using Nedlloyd wished to move to an outsider.	Discounts only where shipper solely using P&O wished to move to an outsider.
	Price per TEU	US$	Slightly below the non structured level	1035 (TAA fixed unstructured level)	1085	1085 (TAA fixed structured level)	1085 (TAA fixed structured level)	1085 (TAA fixed structured level)
	Market segmentation	Yes/No	Yes	Yes	Yes	Yes	Yes	Yes
		Details	On a market basis - by commodity, by customer, by service quality.	Traditional segmentation as allowed by TAA	Traditional segmentation as allowed by TAA	Traditional segmentation as allowed by TAA	Traditional segmentation as allowed by TAA	Traditional segmentation as allowed by TAA
	Subsidy	Yes/No	No	No	Yes - indirect	No	No	No
		Amount	N/A	N/A	Unknown	N/A	N/A	N/A
		Type/details	N/A	N/A	Indirect . Preferred German carrier for military goods, state employees and goods.	N/A	N/A	N/A

Figure 5 European Community based competitor's framework

166

PRICE	Currency convertibility	Yes/No/Partial		Yes	Yes	Yes	Yes	Yes	Yes
		Acceptable currencies		All convertible currencies	All convertible currencies	All convertible currencies	All convertible currencies	All convertible currencies	All convertible currencies
	State control	Yes/No		No	DSR Yes, Senator No	No	No	No	No
		Details		Italian private ownership	DSR previously owned by DDR, transformed to German state ownership on reunification	State owned but greater independence, eg prices, income, staff, vessels. Joint stocked into separate companies ready for privatisation	Part of AP Moller Group	Rotterdam based company amalgamated from 4 Dutch based shipping companies in 1970	UK registered company listed on stock exchange, no state ownership
	Profit/loss	US$		10 million profit		75 million loss	Moller pre tax profit $370m	Loss Fls 58.2m	£36.2m profit (container and bulk)
	Competition	Details		Both national and international	Both national and international	Both national and international	Both national and international	Both national and international	Both national and international
PROMOTION	Sponsorship and hospitality	Details		No sponsorship, hospitality locally decided eg golf	None	Limited amount	None	Limited	Occasional in UK eg. Grand National, Ballet, Jetski team
	Advertising	Yes/ No		Yes	Yes	Yes	Yes	Yes	Yes
		Budget		Recently reduced	Unknown	Almost zero, reduced from Dmn in 1991	Head office determined	Limited	Limited
		Press used	Gen.	No	Very occasional eg Journal of Commerce	Very occasional	Occasional - related to special issues	Occasional	Occasional - related to special issues
			Ship.	Yes eg Lloyds Loading List	Yes, eg De Lloyd, Lloyds Loading List	Occasional	Yes eg Lloyds Loading List	Yes eg Lloyds List, IFW	Yes eg Lloyds Loading List
		Frequency		According to publication or occasional	Regularly in local press	Infrequent	According to publication	According to publication	According to publication

Figure 5 European Community based Competitor's framework - continued

P R O M O T I O N

Category	Type	Info.							
Advertising		Permuasion	Recently reduced	Unknown	Almost zero, reduced from Dmn in 1991	Head office determined	Occasional	Limited	Limited
		Presence	No	Very occasional eg Journal of Commerce	Very occasional	Occasional - related to special issues	Occasional	Occasional	Occasional - related to special issues
Image		Corporate image Yes/no	Yes eg Lloyds Loading List	Yes, eg De Lloyd, Lloyds Loading List	Occasional	Yes eg Lloyds Loading List	Yes eg Lloyds List, IFW	Yes eg Lloyds List, IFW	Yes eg Lloyds Loading List
		Details	According to publication or occasional	Regularly in local press	Infrequent	According to publication	According to publication	According to publication	According to publication
Representatives		Yes/No	Yes, eg. Lloyds Loading List	Yes	Yes	Yes eg Lloyds Loading List	Yes	Yes	Yes eg Lloyds Loading List
		Number	No	No	No	No	No	No	No
		Locations	Very important	Yes eg International Freighting Weekly	No	Yes	Yes eg Financial Times	Yes eg Financial Times	Yes eg Lloyds List, Financial Times
	Agents	Yes/No	Yes	Yes	Yes	Yes	Yes	Yes	Yes
		Number	Consistent and comprehensive	Consistent and comprehensive	Consistent and comprehensive	Consistent and comprehensive	Consistent and comprehensive	Consistent and comprehensive	Consistent and comprehensive
		Location	Yes	Yes	Yes - company policy to use regionally owned companies	Yes	Yes	Yes	Yes

Figure 5 European Community based Competitor's framework - continued

		Col 1	Col 2	Col 3	Col 4	Col 5	Col 6
Market research	Yes/No	Virtually none	Yes	No - indirect from other activities	Yes	Yes	Yes
	Budget	None specific	Agent based	0	Centrally controlled	undisclosed	undisclosed
	Number employed in marketing department	Trade manager for each route carries out all research	No specific personnel - through customer contact	0	In house operations in each of 195 locations	Around 10% of employees, 5 specifically in UK	Integral part of operations, internal except in USA
Public relations (PR)	Budget	Reduced	Part DSR Senator funded Part local agency funded	0	Centrally controlled	Limited	Limited
	Promo-tional items — Types	Generally inexpensive items eg. Calenders, pens, diaries.	Sweaters, liquors, paper blocks, ties, lighters etc	Calendars	High quality items form corporate brochures eg cuff links, golf balls	Mugs, pens, pencils as giveaways, plus quality gifts eg bottled ships	Quality pens, calendars etc
	Promo-tional items — No.s	Sufficient for blanket coverage	Plenty	Very few - specific customers only	Limited and strategically targeted	Limited	Limited
Directories	Details	All major worldwide shipping directories, including Lloyds, Fairplay, Containerisation International etc	All major worldwide shipping directories, including Lloyds, Fairplay, Containerisation International etc	All major worldwide shipping directories, including Lloyds, Fairplay, Containerisation International etc	All major worldwide shipping directories, including Lloyds, Fairplay, Containerisation International etc	All major worldwide shipping directories, including Lloyds, Fairplay, Containerisation International etc	All major worldwide shipping directories, including Lloyds, Fairplay, Containerisation International etc
Literature	Yes/No	Yes	Yes	Yes	Yes - extensive range	Yes	Yes
	Types — Brochure	Yes	Yes eg reports	Yes eg reports and accounts	Yes - including regional and corporate	Yes - trades and corporate	Yes - wide range eg container listings
	Types — Booklet	Yes	Yes eg schedules, agents booklets	Yes eg schedules	Yes eg schedules	Yes eg schedules	Yes - more limited
	Destinations	Commercially distributed on demand	Commercially distributed on demand	Commercially distributed on demand	Commercially distributed on demand	Commercially distributed on demand	Commercially distributed on demand
	Languages	English	English	German and English	English	English	English

Figure 5 European Community based competitor's framework – continued

Section	Category	Subcategory						
PROMOTION	Agencies	Yes/No	No	No	No	No	No	Yes - in USA only
		Types — PR	N/A	N/A	N/A	N/A	N/A	Yes - USA
		Types — Advertising	N/A	N/A	N/A	N/A	N/A	Yes - USA
		Budget	0	0	0	0	0	Unknown
	Exhibitions	Yes/No	Yes	No	No	Yes	Yes	Yes
		Numbers	Limited	N/A	N/A	Limited	Limited	Limited
		Location/details	International, as appropriate	N/A	N/A	International as appropriate	International as appropriate	International as appropriate
	Press releases	Yes/No	Yes	Yes	Yes	Yes	Yes	Yes
		Numbers	Variable - as circumstances require	As circumstances require	As circumstances require	Variable - as circumstances require	Variable - as circumstances require	Variable - as circumstances require
		Languages	English	English	English	English	English	English
PHYSICAL EVIDENCE	Containers	Numbers owned and leased	4,800	18,057 owned, Leased 53,977	84,394 owned, 3,585 leased	Over 125,000	110,000	127,000 (84,000 owned, 43,000 leased)
		Numbers in own colours	All owned containers	All owned and long leased	All owned and long lease containers ie 85,981	All	All	All
	Vehicles	Number/type of lorries	Own trucking company - Logoped	0 - third party trucking	0 - all subcontracted	Own various trucking companies	Trucking companies in Europe	Trucking companies in Europe
		Number/type of others	Rail and terminal equipment	0	0	None	Rail wagons, aeroplanes, offshore drilling etc	Limited port equipment

Figure 5 European Community based competitors' framework - continued

170

P H Y S I C A L E V I D E N C E	Company Offices — Number	9	7	28	196	76 own offices, 140 agents offices	232
	Company Offices — Details	Head Quarters in Ipswich, offices in UK, Belgium France, Germany, Holland, Italy, Spain, Switzerland	Germany, Cyprus, Dubai, Hong Kong, USA	Austria, Belgium, Czech Republic, France, Germany, GB, Hungary, Italy, Netherlands, Poland, Switzerland.	HQ Copenhagen, 195 regional offices	HQ Rotterdam, Offices in Europe, USA, Canada, Central and South America, Caribbean, Middle East, Africa, Asia, Australia, Pacific.	HQ London, Main Offices in Rotterdam, Singapore, Bombay, UAE, Sydney, Wellington NZ, Jeddah, New Jersey.
	Ports served as part of North Atlantic service — Number served	7	12	15	11	11	13
	Ports served as part of North Atlantic service — Locations	La Spezia, Rotterdam, Felixstowe, Le Havre, New York, Norfolk, Savannah	Le Havre, Rotterdam, Bremerhaven, Hamburg, Felixstowe, Antwerp, Oakland, Long Beach, Cristobal, Savannah, Norfolk, New York	Le Havre, Antwerp, Hamburg, Bremerhaven, Gothenburg, Rotterdam, Thamesport, Liverpool, New York, Portland Maine, Boston Massechussets, Baltimore, Hampton Roads - Norfolk and Portsmouth - Savannah	Le Havre, Felixstowe, Rotterdam, Bremerhaven, Halifax, Boston, New York, Norfolk, Charleston, Port Everglades, Houston	Hamburg, Bremerhaven, Antwerp, Felixstowe, Rotterdam, Le Havre, New York, Boston, Norfolk, Halifax, Charleston	Bremerhaven, Felixstowe, Le Havre, Rotterdam, Boston, New York, Norfolk, Charleston, Port Everglades, Houston, Long Beach, Oakland, Jacksonville.
	Routes served regularly — Number and details	Over 100 ports of call on 17 liner services involving: Europe - Australia, Far East. North Europe - Mediterranean. Round the World. Mediterranean - US Gulf. Middle East - India.	Round the World - both directions - ie Europe, Middle East, SE Asia, Far East, USA, Mediterranean, Middle East, South East Asia, Far East.	Atlantic - from US East coast and Canada, US/Canada West coast - Europe. Asia/ North America, Caribbean, Venezuela/Columbia, US Gulf/Mexico, Central/South American West coast, Brazil, Madagascar, Jeddah/ Arabian Persian Gulf, India/Pakistan, Australia/New Zealand, China, Red Sea, Far East.	Europe - Far East/ Middle East/ Mediterranean/ West Africa/ USA/ Canada/ S America West coast, Inter European	Over 30 liner services operating world wide	Far East and Jeddah, N America, USA/Middle East, Gulf, Caribbean basin, S, W central and E Africa and Red Sea Region, Australia, NZ, E Mediterranean, Middle East Gulf and Asia, Bay of Bengal.

Figure 5 European Community based competitors' framework - continued

PROCESS								
Conference and Consortia	Membership Yes/No	No	Yes	Yes	Yes	Yes	Yes
	Title	N/A	TAA	TAA	TAA	TAA	TAA
	Level of Membership	N/A	Non structured	Unstructured member	Structured	Structured	Structured
Guaranteed traffic	Yes/No	No	No	No	No	No	No
	Details	N/A	N/A	Some preferred traffic	N/A	N/A	N/A
Collaboration	with road	Own trucking company -Logoeped	Commercial relationship with independent contractors	Yes - commercial relationships with independent contractors	Own trucking companies	Own companies	Own companies
	with rail	Two Italian rail freight companies - Sagomar and Spedrans	Commercial relationships with independent contractors	Yes - commercial relationships with independent contractors	Commercial contracts	Wagon ownership	No
	others/details	Contrepair. Major terminal operator eg. La Spezia run by Contship Italia	Own logistics company ETL for European truck and container control, also tramp and passenger ferry activities	No details	Tankers, Offshore supply, bulk, car carriers, drilling off shore, shipbuilding oil and gas carriers, airlines, EDP, car components, plastics, medical services, electronics, PVC, supermarkets	Logistics, warehousing, air freight, trucking	Warehousing and logistics in house companies

Figure 5 European Community based competitors' framework - continued

172

			see physical evidence for details	see physical evidence for details	see physical evidence for details	see physical evidence for details	see physical evidence for details	see physical evidence for details
P L A C E	Origins and destinations served	No.s and ports served						
	Computer Reservation Systems	Yes/No	Yes	Yes	Yes	Yes	Yes	Yes
		Details	Every office	Every office	Every office	Every office connected by satellite computer system using in house EDP company to provide full management information systems	Every office	Every office
P E O P L E	Qualifications and training	Details	No management training scheme, technical skills come first, management skills acquired. Some IT training, minor allowances for accountancy. Little time, encouragement or finance available.	No evidence of in house training schemes, under German law the company is required to run graduate training schemes for new employees only	Shore based management training schemes. Mainly IT,ship management and general; management. Mainly in house with some occasional use of outside organisations eg Harvard	Internal and external training at HQ, range of courses eg Global sales training involves seven schemes of 3 days each. Training used as motivational issue	In and out house training for IT, technology, product development and customer service.	Graduate recruitment scheme for management. In house training for lower to senior management, left to individuals to take initiative.
	Employees	Number at sea	Only total number available	Only total number available	700	Only total number available	800	1000
		Number ashore	Only total number available	Only total number available	3,450	Only total number available	3200	4000
		Total	900	3,500 approximately	4,150	25,00 for Moller group	4000 (25,000 worldwide in Group)	5000
	Management Structure	Details	Various line managers responsible to MD, subsequently responsible to Owner	Directors for each region are controlled by an Executive board consisting of one chairman and three directors - one each for finance, Eastern hemisphere and Western hemisphere.	Four MDs - one each for Europe, West, East, and Fleet/Containers - report to Executive Director in Hamburg	Maersk Line as a division of AP Moller Group is split into regional offices - established according to Danish law	"Traditional pyramid" style management structure	P & O Steam Navigating Company is split into various sections, eg Containers. Each section is run by a Managing Director who controls directors for each trade.

Figure 5 European Community based competitors' framework

		None	Senator 0, DSR some	Shore - minority Sea - majority	Minimal	Regionally varied	Minimal
PEOPLE Unions/Workers Councils		None	Senator 0, DSR some	Shore - minority Sea - majority	Minimal	Regionally varied	Minimal
	Details	Only legally required union membership	Senator has no union membership, whilst DSR still has some due to tradition of the old regime	Sea farers organised through workers councils	Unions tolerated, staff paid well to overcome labour problems	Locally determined eg Netherlands have strict regulations for both shore and sea based staff.	Unions only where legal requirement, eg ports in some countries. No shore based recognised union.
PRODUCT Speed of North Atlantic service	Vessel Speed (kts)	Variable - see 'ship types'	18	20.5	23 or 24	19	19
	Days (Felixstowe to New York)	11 or 12	13	London to New York = 9 days	10	11	11
Safety	Containers lost p.a.	No data available	No data available	No data available	No data available	No data available	No data available
Reliability	Number of late deliveries p.a.	No data available	No data available	No data available	No data available	No data available	No data available
Frequency of North Atlantic service	Number of services per week	One service every 18 days, west bound only	One weekly fixed day service in each direction	Two	3	1	4
Quantity	TEU capacity on North Atlantic	Approx. 6,000 carried	Approx. 125,000 carried	707,000 TEU actually carried	Approx. 150,000 carried	Approx 125,000 carried	3,456 TEU capacity per vessel
Ship types	Number of each type (see appendix 4 for abbreviations)	Variable - All ships chartered in with crew and officers	24 vessels of between 2700 and 2850 TEU (including 8 owned by Cho Yang) are operated as the Tricon service	28 vessels including 18 container vessels	55 vessels including gas tankers; support, supply, container and ro ro vessels; crude, product and car carriers.	59 vessels including multi purpose, container and ro ro types	34 container vessels ranging from 14,055 dwt to 58,600 dwt

Figure 5 European Community based competitors' framework

174

P R O D U C T	**Ship Ages**	Average years	Variable - see above		7.7	8.5	9.5	Mainly Hong Kong, Gibraltar, - ie crown dependencies unless chartered
	Flag	Details	Largely German - newbuilds from German yards, German officers, Philippino crew	DSR vessels all German, Senator Liberian or Panamanian	Majority German, 2 planned to flag out to Singapore	Largely German, using German officers and Philippino crew	Dutch, Surinamese crew or multinational crew	Mainly Hong Kong, Gibraltar, - ie crown dependencies unless chartered
	Cross trades	Yes/No	Yes	Yes	Yes	Yes	Yes	Yes
	Computer tracking for containers	Details	Yes	Yes	Yes	Yes	Yes	Yes
	Trading terms	Details	According to contract	According to contract	According to contract	According to contract	According to contract	According to contract

Figure 5 European Community based competitors' framework – continued

4. Market segmentation

The companies studied all segmented the market on standard commercial bases, for example by customer or by commodity, within the confines allowed by the TAA. The exception here is Contship who provide independent competition due to their freedom of segmentation.

5. Subsidy

As a general rule none of the lines received any subsidy by the end of 1992. The exception here is Hapag Lloyd who admitted receiving some indirect subsidy and were the preferred carrier for state sensitive goods, and state employees. Fairplay quoted a figure of DM 50m as a total of the amount of subsidy provided to German shipping during 1992, although the amount received by Hapag Lloyd remains undisclosed (Fairplay 22/10/92) . DSR received direct and indirect subsidies under the old regime, but these ceased upon German reunification in 1991. Overall the position of the competitors is of independent companies operating in a free market with little or no subsidy.

6. Currency convertibility

Each of the European Community companies are based in countries with fully convertible currencies and operationally they will accept any convertible currency as payment.

7. State control

Notably DSR, formerly owned by DDR was transferred to German state ownership upon reunification in 1991. The position for all other Community based operators is one of private ownership with no direct state involvement. However, there is some evidence of government influence and the effects of politics upon operators; witness Maersk who are Denmark's third biggest export carrier.

8. Profits

Of the figures available for this framework four companies made outright losses on all shipping activities during 1992, and two were profitable:

Losses:

Nedlloyd	Fls	58.2m
Hapag Lloyd Liner Division	DM	50 m
Senator	DM	30 m

DSR	DM	19.7m
Profits:		
P&O (Container and Bulk shipping)	£	36.2m
Contship	US$	10 m

Many of the losses made by the Community competitors can be attributed to poor returns in the liner market at that time, typified by those on the North Atlantic. It also reflects the highly competitive and rather curious structure of the North Atlantic market - whereby companies continued to underprice their services - thus making losses - whilst continuing to offer a high service level. This defies conventional market wisdom - which suggests that a reaction in cost or service level should have occurred - and provides the basis for the development and introduction of the TAA.

Promotion

1. Sponsorship and hospitality

Due to financial limitations the companies studied generally carried out little (eg Contship) or no (eg DSR-Senator and Maersk) sponsorship or hospitality during 1992. There was however greater evidence of these activities at P&O who sponsored the Royal Ballet and the UK Jetski team and provided hospitality at the Grand National and golfing events. Overall the low level of activity in these areas can be seen as a reflection of low returns in the liner market creating financial limitations.

2. Advertising

The majority of companies studied carried out a relatively small amount of advertising in late 1992 due to financial limitations. An outstanding example of this is the reduction of Hapag Lloyd's advertising budget from around DM 8m in 1991 to almost zero in 1992. However, all companies, including Hapag Lloyd still advertised in the shipping and related press, for example Lloyds Loading List, and the occasional Lloyds List, International Freighting Weekly, Containerisation International and the Journal of Commerce. The general press was used very occasionally, and was usually linked to special issues. The advertising aims of all companies were to maintain presence in the market, and to provide information about their company and services. The overall position of European Community companies at the end of 1992 was that their financial position only permitted a minimal amount of advertising at the time that they needed to increase business to survive.

3. Image

Each of the European Community based companies had a clear and comprehensive corporate image. The 'images' were usually created using various styles, colours and logos, and they had to appear in every promotional activity. The aim of each company was to convey a consistent image of quality, which was readily identifiable, and which reflected the way that they wished to be perceived.

4. Representatives and agents

All the companies questioned employed both representatives and agents to cover as wide a scope as possible in their relevant areas of the world. The feeling amongst most companies was that owned agents and representatives were the better option. This was true at Maersk (195), and particularly at P&O (228) who owned all representatives and agents where possible, ie everywhere but Canada or Eire. The exception to the rule was Hapag Lloyd whose policy was to use regionally owned but independent companies.

5. Market research

No specific budget or personnel were employed for market research purposes by Hapag Lloyd, DSR-Senator or Contship. Market research here was carried out internally by trade managers or representatives. The other three companies carried out this activity in specific in house areas, with Nedlloyd claiming to have around 10% of its staff working on market research at any one time. Again, low levels of market research can to some extent be attributed to financial limitations.

6. Public relations (PR)

Limited budgets have placed financial restraints upon PR activities especially at Hapag Lloyd where there was virtually no budget available for PR. However, all the companies studied provided promotional items in late 1992/ early 1993, such as pens and calendars. Although many referred to these items as 'throwaways' they were usually produced to the highest possible standard and were in keeping with the corporate logos to further promote the desired corporate images of quality. More expensive gifts, such as sweaters, glass bottled ships, and even liqueurs, were usually reserved as loyalty bonuses for established customers, through a process of strategic targeting.

7. Directories

In early 1993 all the companies surveyed appeared in major shipping directories with world wide coverage, for example Lloyds Maritime Directory, Fairplay, and

Containerisation International. This would appear to be standard practice throughout the shipping world.

8. Literature

All the companies in the study produced literature in early 1993. This was usually in the form of corporate brochures, schedules, annual reports and accounts. Maersk produced one of the most extensive ranges which included fleet, reefer and regional brochures. Literature is generally distributed on commercial and demand bases. All companies produced their literature in English - shipping's international language - with Hapag Lloyd also supplying German literature. The overall position of the Community based companies was one of high quality literature, with strong corporate images, reflecting quality market positions.

9. Agencies

The study revealed that advertising agents were rarely used by shipping companies, which reflected the general nature of the industry. Of the selected European Community based companies the only one to use advertising agents were P&O and this only applied to their activities in the United States, and hence was for market related reasons rather than company preference.

10. Exhibitioneering

Five of the six companies questioned stated that they did not carry out any exhibitioneering during 1992, mainly due to lack of financial resources. The only exception is Contship, where a limited amount of exhibitioneering was carried out - mainly by the parent company.

11. Press releases

All companies carried out press releases during 1992 as circumstances required. No specific examples were given, but press releases are standard shipping practice, and are usually produced in English, shipping's international language.

Physical evidence

1. Containers

The companies included in the framework all owned and/or leased substantial numbers of containers, including 85,000 at Hapag Lloyd, 185,000 TEUs at P&O to 110,000 of 27 types and eight sizes at Nedlloyd.

For each of the companies studied the majority of containers are usually owned, with some long lease and some short lease to cover peaks in the market. All owned and long leased containers were painted in the company's own colours to maintain consistency of corporate images.

2. Vehicles

The majority of companies owned or had associated trucking companies, although Hapag Lloyd and DSR-Senator used third party subcontractors. Other vehicle ownership ranged from none at DSR-Senator, Hapag Lloyd, and Maersk, to P&O owning a limited amount of port equipment, and Nedlloyd operating a wide range including rail wagons, aeroplane subsidiaries and offshore drilling. Where vehicle ownership was limited the European Community based shipping companies usually used those belonging to associated companies or those of companies with which they had formed strong collaborative and commercial links.

3. Company offices

There is no standard pattern of company offices; each situation varied according to company policy and size. Office ownership ranged from DSR-Senator who have seven offices in Germany, Cyprus, Dubai, Hong Kong and USA, to Maersk with 196 offices spread throughout the world.

4. Ports served regularly as part of North Atlantic service

Of the seven companies listed all called at New York, six called at Le Havre, Rotterdam, and Norfolk, and five visited Bremerhaven. The range of other ports visited as part of the service varied considerably (see figure 5) from Hapag Lloyd serving fourteen ports as part of this service to Contship serving seven.

5. Routes served regularly

The companies listed offered a range of services, with no real pattern. Examples included:

Nedlloyd -   30 liner services world wide
DSR-Senator -   RTW (both directions), Mediterranean, Middle East, South
                  East Asia, Far East
Maersk -   From Europe to Far East, Middle East, Mediterranean/West
                  Africa, USA/Canada, South America/West Coast, and inter
                  European services.

The full extent of the European Community operators services are shown in

Figure 12.

Process

1. Conference and consortia

Of the six companies studied, five belonged to the TAA. P&O, Nedlloyd, Maersk and Hapag Lloyd were all structured members of the TAA, whilst DSR-Senator was an unstructured member. Contship remained an independent company.

2. Guaranteed traffic

None of the companies questioned claimed to receive any guaranteed traffic. However, Hapag Lloyd were the preferred carrier for state sensitive goods and state employees.

3. Collaboration

Four of the companies listed in the framework used their own road haulage companies, these being P&O, Nedlloyd, Contship, and Maersk. Hapag Lloyd and DSR-Senator used independent contractors. For rail transport, commercially based contracts with third parties were more usual, although Nedlloyd and Contship used their own wagons and companies respectively. Other collaboration involved warehousing and logistics companies which were occasionally in house.

Place

1. Origins and destinations - see physical evidence

2. Computer reservation systems

Computer reservation systems were widely used by all six Community based companies. Perhaps the most impressive and extensive example of this was provided by Maersk. At Maersk all offices were connected by a satellite computer system. They used an in house EDP company to provide full management information systems, and to provide details of shipments, among a wide range of highly varied facilities.

People

1. Qualifications and training schemes

The majority of companies used in house training schemes with only occasional use of outside organisations. Only DSR had no evident in house training schemes, although under German law the company was required to run graduate training schemes for new employees. At the other end of the scale Maersk ran both internal and external schemes at their head quarters in Copenhagen. Training was used as a motivational issue with a range of courses offered; for example global sales training ran as six or seven schemes of two to three days each. The high profile of training reflected their position as a quality operator.

2. Employees

The number of employees varied widely. Contship employed a total of 900 people, while DSR-Senator and P&O Containers employed between 4000 and 5000. Nedlloyd's total employee figure was in the region of 25,000 - although this included a wide range of non shipping activities.

3. Management structures

The general pattern within the shipping companies listed was for a 'pyramid style' management structure. There was often an overall Executive Board under which a Managing Director oversaw Directors for each particular trade. Contship and P&O provide two examples:

Contship

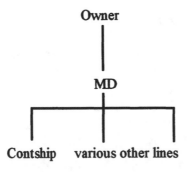

P&O

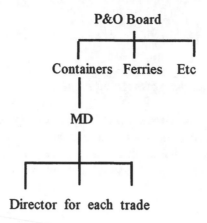

4. *Unions*

Union membership was usually regionally based. On the shore side union membership was often limited or not recognised at all, such as at P&O. Sea staff were more likely to belong to a union, such as the majority of Hapag Lloyd's seafarers. The main exception was DSR where due to the tradition of the old regime unions remained fairly strong - this was in sharp contrast to the situation of many shipping companies still situated within the Eastern bloc, such as POL, where union power had become diminished due to its association with the old regime.

Product

1. *Speed*

The speed of service offered by each company varied according to a wide range of factors including routes taken, type of service, vessels used, fuel costs and company policies. It would therefore be misrepresentative to compare the days taken by different companies to reach different ports. Generally, the time to reach New York from Felixstowe was between 10 and 12 days, and to reach Boston was around 9 days.

2. *Safety*

Due to the commercial sensitivity of this data, no safety nor container loss records were made available to the research, although each company questioned claimed to have a good safety record.

3. Reliability

Again no data was available for this part of the research due to commercial reasons.

4. Frequency of services on the North Atlantic

The range of frequencies of services reflected the wide scope of services offered by the six Community based companies. These included four sailings a week by P&O from Europe to North America and back, a weekly service by Maersk, and DSR-Senator, and a service every 18 days westbound only by Contship.

5. Quantity

This proved a difficult figure to obtain for the specific market but some examples were provided by Nedlloyd who moved approximately 125,000 TEUs across the North Atlantic during 1992, Maersk who moved approximately 150,000 TEUs, and Hapag Lloyd who moved a total of approximately 210,000 TEUs east and west bound over the same period. No separate figures were available for west bound alone.

6. Ship types

The number and type of vessels operated varied between the companies. For example DSR-Senator operated 24 vessels, whilst Maersk operated 43 vessels. Over the same period Hapag Lloyd operated 18 container vessels, 1 passenger vessel and a number of smaller vessels including tugs and launches, whilst P&O, amongst a fairly extensive fleet, ran 28 purpose built container ships. The number and type varied according to company size, policy and markets served, but all vessels carried their respective company's corporate image.

7. Flag

Flag varied according to the country in which the company was based and the company policies. Flags used in 1992 were as follow:

Maersk	Danish
Contship	mainly German
Hapag Lloyd	mainly German
Nedlloyd	Dutch
DSR	German
Senator	Liberian, Panama

Again these results indicated a position of quality, with very little use of the so called 'flags of convenience'.   It remains to be seen whether this position will change as financial pressures encourage companies to flag out.

8. Cross trades

All companies participated in the cross trades both as part of the activities upon the North Atlantic and elsewhere. This is standard shipping practice, and is to be expected in such an international industry.

9. Computer tracking for containers

Of the six companies studied, all used computer tracking for containers as standard.

10. Trading terms

Trading terms were usually based upon contract and were in accordance with current market conditions.

The framework has provided us with an overall picture of the European Community based competitors' position. Summarising each element in turn:

Price

Each of the six companies were in a similar situation as regarded price. The price element was mostly controlled by the TAA, for example in terms of level of charges, rebates and discounts. They each were struggling to make profits in a depressed market and each received little or no subsidy. However, generally it can be said that the European Community operators were aiming at the quality end of the market through their choice of TAA membership, and therefore were able to charge the associated higher prices.

Promotion

It is paradoxical to note that promotion was often limited due to financial restraints, at a time when companies needed to increase business. However, those promotional activities which were carried out were of notably high quality, for example the high standard of PR items produced. All promotional activities were geared towards creating a clear and consistent quality image - reflecting the position the companies wished to occupy within the market.

Physical evidence

Generally, the European Community operators attempted to create a high profile, through quality physical evidence and consistent and comprehensive company image. Examples of this include painting all owned and long leased containers in the company's own colours, and where appropriate maintaining a presence on routes of global importance, such as the North Atlantic, even if this meant incurring losses.

Process

Overall the companies were involved in significant process activities. Firstly, the majority - five out of six - belonged to the TAA. Secondly, all were involved in a strong degree of collaboration either through ownership of related companies - for example P&O and Contship owning road haulage companies - or through established commercial relationships - which all companies had established with rail, logistics and warehousing interests.

Place

The place element for the majority of the companies was also strong. For example computer reservation systems were used without exception, with Maersk providing the best example, and extensive ranges of origins and destinations were offered. These examples further reflect the required position of quality operators, with a strong position within the market place.

People

The quality position occupied by the European Community based competitors was enhanced by the issues relating to the people element. These issues included low levels of union power, but were perhaps best exemplified by the degree of importance placed upon issues such as qualifications, training and motivation, particularly evident at Maersk.

Product

Generally the product offered was one of quality, and most of the product elements reflected this, with examples existing in terms of speed, frequency and efficiency. A number of companies, such as P&O, had impressive fleets, with the majority sailing under 'respectable' flags. Again the product element pointed towards quality conscious companies operating at the upper end of the market.

Overall it can be said that the European Community competitors generally offered a high profile, quality service for which they could charge a higher price, and that consequently they occupied a relatively strong market position on every aspect - emphasised through their TAA membership and collaboration.

Having looked at each element of the framework in some detail we will go on in chapter eight to compare the differences in position between POL and its European Community based competitors on the North Atlantic trade, and in particular whether POL changed its position relative to the European Community operators during the 1988 - 1992/3 period.

8 Positional comparison of shipping markets

In the previous two chapters we have established the general market positions for Polish Ocean Lines and its European competitors on the North Atlantic trade. The frameworks created in these chapters have provided a structure (using the 7 P's and their elements to assess each) which will enable us to go on to compare the relative market positions, to assess whether POL has repositioned in any aspect. Earlier discussions have shown that it is widely agreed that POL's position is seen as some way behind that of western operators - for example in over employment, dated management structures, promotional structures and attitudes towards profit making and privatisation - views clearly backed by Containerisation International (1993), and Ernst and Young (1990). We will now also attempt to establish whether this position in the market place is stable or changing - that is, whether POL is moving towards the European competitors' position, attempting to do this, falling further behind or simply changing direction between the years of 1988 and 1992/1993.

This will be achieved using the matrix approach described in chapter three. As no suitable technique exists to enable positional comparison the method has been developed by the author. The use of matrices is supported throughout the thesis as a useful and useable display technique (Patton 1987, Friend and Hickling 1987), and consistency is further maintained by the use of this qualitative rather than quantitative analysis. By setting out each of the positions on the seven P's for POL on one axis, and the seven P's for the European competitors on the other, we will be able to compare and assess every possible combination of positional movement. The matrix will examine the 1988 and 1992 positions for POL, and the 1992 position for the Community based competitors. This should indicate whether the positional changes which have taken place in POL have resulted in movement towards or away from the Community operators aggregated position. POL's movements toward or away from the 1992 EC position are represented by arrows. It must be emphasised that this technique only attempts to create a visual display of directional movement, there is no attempt at quantification. Although the comparison is difficult because it attempts to compare a diverse range of information, it is essential to assessing any positional movement by POL. The complete matrix is shown in figure 6 - spilt into four sections for display purposes - and it is to this that we will now turn our attention by comparing POL's situation for each of the seven P's in 1988 and 1992 with that of the Community based competitors in 1992:

Polish Ocean Lines' price

In broad terms the 1988 position on price for POL can be described as low. At that time POL offered a market led price which was below that of many other operators, and which reflected a cheaper, lower quality service. By 1992 POL had become an unstructured member of the TAA and had raised its price accordingly. This movement of raising price and bringing pricing structures directly into line with European operators can be seen as representing part of a rise in service quality generally.

By comparison the position of the Community operators appears high in each of the seven P's. The competitors charge a high, TAA controlled price for a highly marketable quality service product. They undertake significant and sophisticated promotional activity to maintain a quality image. They possess significant levels of quality physical evidence. Process is a prominent aspect with considerable levels of collaboration and conference membership. They employ high levels of qualified people and promote training schemes, and they occupy distinct, strong and favoured positions within the market place.

Here, it may be concluded that by a positional change in terms of price, POL has moved directly towards the quality position shown by each of the seven P's of the Community based operators. This can clearly be seen in figure 6 by looking at the direction of the arrows indicating positional movement within those cells which relate to POL's price element.

Polish Ocean Lines' promotion

In 1988 the position of POL in terms of promotion was low both in terms of quality and quantity. By 1992 this position remained unchanged. Despite plans to raise the level of promotional activities in terms of quantity and quality, POL's ambitions were thwarted by financial problems.

Where the promotional element of POL is compared with the competitors, figure 6 again shows that the Community based operators remain high in each of the seven P's, as discussed above.

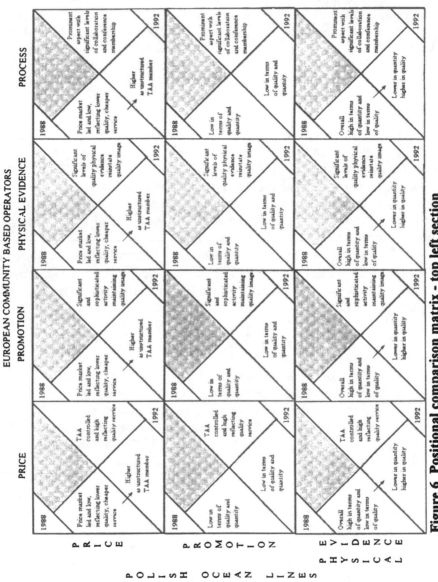

EUROPEAN COMMUNITY BASED OPERATORS

Figure 6 Positional comparison matrix – top left section

190

EUROPEAN COMMUNITY BASED OPERATORS

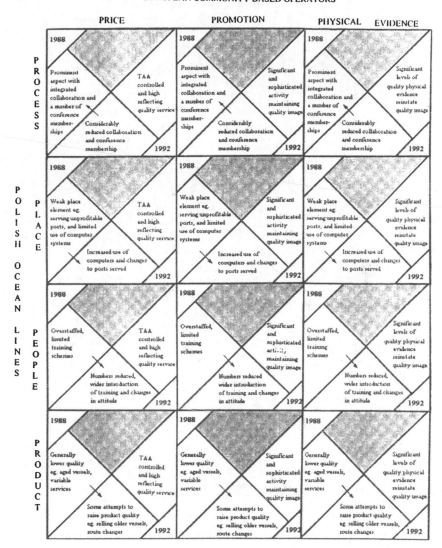

Figure 6 Positional comparison matrix - bottom left section

191

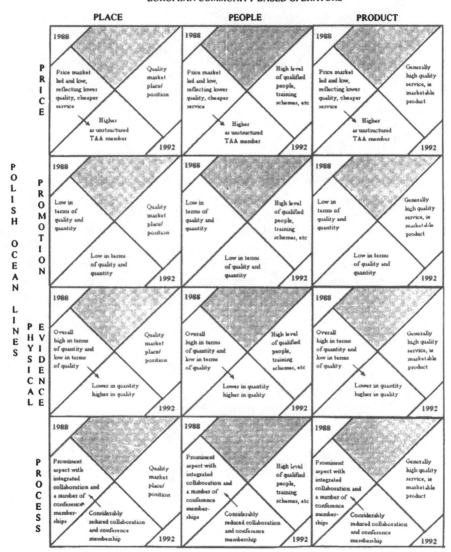

Figure 6 Positional comparison matrix - top right section

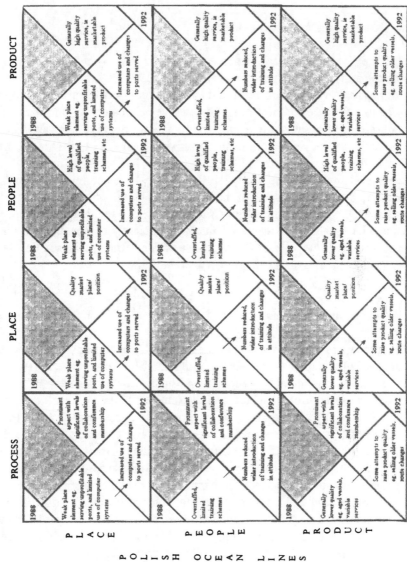

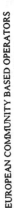

Figure 6 Positional comparison matrix - bottom right section

The consistently low promotional position of POL and the high quality position of the competitors in each of the P's therefore remain mismatched. The lack of any arrows in this section of the figure reiterates the point the there has been no positional change by POL in promotional terms.

Polish Ocean Lines' physical evidence

In 1988 the level of physical evidence utilised by POL was high in terms of quantity and rather low in terms of quality. By late 1992 POL had rationalised its level of physical evidence, decreasing quantity in an effort to raise quality - for example through route and port reduction, and vessel depletion - and the quality was consequently raised - for example by lowering average ship age and by changes to the routes served, for example removing some ex CMEA ports from regular services.

As mentioned previously, the European Community based competitors' position was one of quality in every aspect of the seven P's. Hence, when we compare the changes in POL with the competitors' situation, we can see that POL's physical evidence adjustments have increased quality sufficiently to constitute a direct movement towards the general Community position. Again, this is indicated by the presence of arrows on the diagram shown in figure 6.

Polish Ocean Lines' process

In 1988 process formed a prominent aspect of POL's operations, with evidence of integrated collaboration and membership to a number of conferences. By 1992 collaboration was considerably reduced as subsidiary companies were being sold off prior to privatisation of the whole, and the removal of state control meant that relationships could be formed on a commercial basis rather than dictated by the state. Notably, overall conference membership was also considerably reduced by 1992 although this was said to be due to the prevailing market situation.

By way of comparison each area of the competitors' seven P's reflected their quality image and position in 1992, for example, process was a prominent aspect with significant levels of collaboration and conference membership in most companies studied. Hence POL's reduction of the process element can be viewed as a direct move away from the position occupied by the competitors. Once more, this is demonstrated by the directional arrows shown in figure 6.

Polish Ocean Lines' people

The general position for POL in terms of people was low for 1988; for example

194

POL was over staffed but unable to make redundancies, and there were very few training schemes. By 1992 the position had changed with POL attempting to deal with people by reducing over employment, improving recruitment, increasing training and qualification levels, reducing union power and updating management structures.

As discussed already, the 1992 position for the European Community based competitors was one of quality in each aspect of the seven P's - as one example the companies selected usually employed high levels of qualified people with extensive training schemes. When comparing the positions it can be seen that POL has moved directly towards the Community position during the four year period. This can clearly be seen on figure 6 by examining the associated arrows.

Polish Ocean Lines' product

In 1988 the general standard of POL's product was low, as is illustrated by the use of ageing vessels and variable services. By 1992 there had been some notable attempts to raise product quality through the selling of older vessels and some route changes, to name two examples, although due to lack of finances this was not completed by late 1992. As already discussed, the 1992 position for the Community based competitors was one of quality in every aspect of the seven P's, one example being that they offered a highly marketable quality product, that being a reliable and efficient service often employing impressive fleets.

Consequently, with POL's attempts at raising product quality , the comparison of positions must point once again to a direct movement by POL towards the stance occupied by the Community competitors. This is clearly shown in figure 6 by the direction of arrows relating to POL's product element.

Polish Ocean Line's place

In 1988 POL could be described as having a relatively weak place element, for example sometimes serving unprofitable CMEA ports, and making little use of computer systems to increase product accessibility. By 1992 the situation had improved in terms of quality, by increasing the use of computer systems and altering the origins and destinations served, including withdrawal from the old CMEA market, and changes in administrative offices and agent locations.

As mentioned previously, the quality image of the Community operators is reflected in each of the seven P's and subsequently the attempt by POL to raise quality in terms of place correlates with a direct movement towards each of the competitors elements. This movement can be seen in figure 6 by the presence of arrows in each cell relating to POL's place element.

It is interesting to note that the majority of comparisons shown in figure 6 indicate some movement by POL between the years 1988 and late 1992. Out of a total forty nine cells, thirty five indicate a definite movement of POL's position directly towards that of its European Community competitors. Only seven cells - those associated with POL's process element - indicate a movement which is directly opposite. The remaining seven, that is, those associated with POL's promotional activities, show no directional movement. Despite these encouraging results for POL, it should be noted that on no occasion has their positional movement actually provided a market location equivalent to that of the European Community. In terms of each 'P' broad conclusions can be drawn from the comparisons and their following discussions.

Price

Mainly due to its membership of the TAA, POL has moved directly towards the position of its competitors on the North Atlantic market in terms of price. The various price elements are controlled by the TAA so that POL's membership makes this movement predictable and inevitable once the decision to enter into the consortium has been taken.

Promotion

POL's level of promotion was low in 1988 compared to its competitors based in the European Community - both in terms of quantity and quality. This remained mainly unchanged by late 1992. Despite plans to improve and expand promotional activities, POL did not have the necessary finance available, and in a limited way promotion may even have been reduced - for example advertising. Promotion remains a significant and sophisticated activity of most Community operators, TAA or otherwise.

Physical evidence

The amount of POL's physical evidence fell substantially between 1988 and late 1992, for example, in terms of numbers of vessels, ownership of subsidiary companies and numbers of ports served. If looking at quantity alone, this is a move away from the late 1992 position of the European Community competitors. However, in terms of quality alone the rationalisation improved physical evidence sufficiently to justify considering the actions as a move towards the competitors' position of late 1992. However despite this, one area of poor and declining physical evidence still remains - that of vessel age and condition.

196

Process

Process is a notable exception to the general pattern. The process element which involves aspects such as collaboration and conference membership, is very prominent in the operations of European Community based competitors, but was reduced considerably by POL between 1988 and late 1992. This was due partly to privatisation plans leading to the setting up of independent subsidiaries, and the removal of forced collaboration. Process is perhaps the only area where the position for POL has moved notably away from the competitor's position, although one positive aspect is POL's membership of the TAA, albeit at unstructured level.

People

The position of the European Community based competitors for late 1992 is one involving a very high level of qualified people. POL is moving towards this position as quickly as company finances and Poland's economic policy will allow. POL was over staffed both in 1988 and late 1992, but was not allowed to make the necessary number of redundancies due to state interference. Changes in POL's company policy were beginning to make a difference to training and qualifications of employees by late 1992. Some evidence of changed management attitude was also apparent.

Product

Between 1988 and late 1992 POL's position on product moved directly towards that of its European Community based competitors. The best example of this is the selling of older vessels as a part of raising of the quality of service - although the fleet remains relatively old and in deteriorating condition. Other product aspects (such as frequency) remained constant.

Place

Between 1988 and late 1992 POL's actions regarding the place element represent a further move towards the position of the European Community based competitors. Examples include increased use of computer systems and changes to the origins and destinations served, bringing them more into line with western operations.

Overall, it is clear that POL is repositioning within the market place, and is attempting to adopt a tack similar to its competitors based in the European Community. Generally, POL is attempting to move away from the old image of a lower quality cheaper service, to a better quality more expensive western style of operation with improved products, rationalised place and attempts at improvements in the use and consideration of people. This repositioning is likely to take some time to achieve fully, and as time progresses the chance of the ultimate company aims changing increases significantly. The reader should also be reminded that this research applies only in the context of the North Atlantic trade where the introduction of the TAA has placed controls upon the market. In terms of this research this means that the lack of a free market prohibits complete free positioning by POL. These points indicate some of the many potential areas for further research covered in the next chapter.

9 Conclusions

This research has proven highly topical; witness for example the continuing changes taking place throughout East Europe generally, the wide reaching effects of German reunification, and the on going debate between the TAA and the Commission of the European Community. Notably, the study has had to embrace a wide range of disciplines, and although this multifaceted feature has made it complex to undertake, it has been necessary to ensure that its application and outcomes are relevant and pertinent to the existing market place and the problems faced by POL. Notable as well is its originality both in its coverage of a new situation and its new applications of the various models. Despite this it can be argued that the research is highly transferable as it may be applied to other areas, such as other trades, companies or countries and still produce meaningful and viable results. For example, an analysis of POL's activities in other markets could provide similar or varying results which would help to indicate whether the chosen North Atlantic trade was a representative case study. The research can further be described as stable. Despite some use of qualitative judgement, were the study to be repeated it is more than likely that broadly the same results would be achieved. As noted before a quantitative approach could have been accused of being spuriously objective.

The objectives of this research were outlined at the outset as:

1 To assess the situation of the Polish shipping industry before and after the changes which have taken place, between 1988 - 1992
2 To analyse the effect of East European change upon Polish shipping and to develop models to identify the major areas of change and the relationship of changes to the shipping industry.
3 To analyse through further model development, the significance of change in Polish shipping when related to European Community based companies operating in a competitive market.

In the main, these objectives have been achieved at various stages of the research, through the use of contextual models, and the framework and two dimensional models which were structured by the marketing mix. However, their achievement has been hampered to some extent due to substantial and fundamental changes taking place throughout East Europe, which in effect moved the goal posts of the research. Examples of these changes include the succession of elections which have taken place in Poland, POL's recent (1993) withdrawal from direct operation in the North Atlantic market, changing politics and instability throughout East Europe, the formation of the TAA, the problems in Yugoslavia and the formation of two new countries - the Czech Republic and Slovakia, in place of the former Czechoslovakia.

In the midst of these changes we have been able to establish a market position for POL for 1988 and late 1992/early 1993. Chapter eight revealed that POL was operating by late 1992 at the lower end of the marketing scale, offering few add on services, a cheap price, lower levels of technology, a poor image and little marketing. On a more positive note POL had rationalised ports and routes served, had lowered the average age of its vessels and was looking to improve qualifications and training.

A number of beneficial steps have been made since 1992 including wider installation of computer systems and further staffing reductions. Cuts to fleet size were continuing as POL drove to reduce overheads and create a viable structure for the future (Lloyds List 27/05/94). POL was reported to have reduced its cost base by $50million in 1993 (Containerisation International 1995). This involved four main factors:

1   joint ventures with foreign companies
2   tighter controls on accounts and invoicing procedures
3   tighter container logistics and better repositioning systems
4   smaller companies proving to be more efficient

By May 1995 POL had made notable progress towards privatisation. It had concentrated its assets around the mother company and had released its operating divisions as quasi-independent daughter companies (Fairplay 1994). The group consisted of subsidiary companies as follow (Containerisation International 1994):

1   Euroafrica Shipping Line - Operates and manages ro ro, semi container and conventional services in the Baltic, North Europe and West Africa.
2   POL-America - Owns 32 vessels also operates semi container service to South America.
3   POL-Levant - Operates and manages ro ro and conventional services to the Mediterranean.
4   POL-FESA - Manages Europe - Far East service.
5   POL-Eureka - Manages transatlantic and Mexican Gulf services.

6   POL-Seal - Operates East Africa breakbulk liner services.
7   Polcontainer - Container logistics and management.
8   Polteam - Container repairs.
9   POL-Supply - Warehouse storage and distribution.
10 Franck & Tobiesen Poland - Road haulage.

In 1994 POL gained the legal status of a fully commercial holding company and applied to the government to be recognised as a fully privatised entity. POL's chief commisioner Kowalski stated that although POL was still fully state owned it was important to establish a private limited company, with the aim in the future of going public. He anticipated that this process could be accomplished by 1998. Whatever the outcome, it is likely that the government will maintain some interest in POL having funded the company's asset base.

As to the future, POL would appear to have accepted the need for long term strategic planning, a relatively new concept for individual companies in East Europe, as this would have fallen under the ambit of the CMEA, with plans being centrally controlled. In earlier chapters we saw that POL was repositioning towards the position of European competitors on the North Atlantic, but as time progresses and market conditions change so will the ultimate goals. There are three possible scenarios which may face POL by the turn of the century:

The first scenario is that POL may catch up with the European based competitors and acheive an equivalent market position to these companies. This seems unlikely within this time scale given POL's starting point and the amount of change needed.

The second scenario is that POL will have exited the market completely. Although POL has had to rationalise services considerably, for example withdrawing from operating its own vessels on the North Atlantic, it seems unlikely that POL will exit liner shipping altogether. It is also questionable whether the government would allow POL to fail in the new market environment. Looking to the near future, Lenczewski, POL's chief financial officer, has suggested that POL's financial problems may mean that the company has to move toward becoming an operator rather than an owner (Fairplay 1994).

Perhaps a more likely scenario would be for POL to continue in a similar vein to the present situation - concentrating upon surviving the changes taking place through crisis management.

Other likely future scenarios include a realignment of operations towards Germany and the west, particularly the European Community, of which eventual Polish membership seems inevitable in the long term.

In this case study of Poland, and Polish Ocean Lines, shipping seems likely to survive whatever the problems, but what is clear is that it will be in a very different form - smaller, more modern and more efficient. Although much remains to be done, the attitude of the industry and government is accepted internationally as correct and progress is being made.

In general terms, the amount of restructuring required by East European shipping in order to survive the political, social and economic changes taking place is substantial. The situation faced by POL can be broadly related to other shipping companies throughout East Europe. The research undertaken is highly transferable and could be applied for example to industries in Romania, Bulgaria or the Czech Rebublic, but the case of POL helps to develop an understanding of the general East European situation.

As well as establishing the position of POL in late 1992, a number of issues stand out from the research, many of which raise questions about the future. Notably, these issues fall into two main categories - commercial and theoretical.

Commercial

One of the most basic questions which must be asked is whether East Europe should remain involved in operating and owning shipping, in particular from an economic viewpoint. Although shipping is an efficient method of earning hard currency, the demise of the CMEA has meant that its defence and insularity are no longer issues, and returns on investments may be greater in other areas. However, a decision to exit from the shipping market is complicated by the need to consider its political acceptability. Following on from this, it is peculiar that shipping operators throughout the world often persist in markets where they have made losses for a successive number of years, not least the North Atlantic, and that they often under market such a major industrial product. Each of these points may provide a basis for further research.

Another of the issues raised concerns the emerging evidence of western practice in East European shipping, for example in areas such as marketing, and quality assurance. It may be argued that some evidence of this has always existed due to the international nature of the shipping industry. Further questions on this point include whether or not POL truly wishes to change, and if so, how much this is possible given the initial levels of understanding. This research has shown that POL is moving towards the EC position, but has not discussed whether this movement was planned or unintentional, welcome or feared.

Privatisation is a major issue for POL. Questions here involve the pace of privatisation, and the levels of achievement. For example, whether privatisation is really taking place when subsidiary companies are sold to the state, or whether this is to convince and encourage western investors. Further, whether selling to the west is a politically acceptable option to the electorate.

Privatisation and state withdrawal of regulation will mean that diversification in the shipping industry and beyond becomes possible and likely, although it remains to be seen whether this will occur, how, or to what extent. Alternatively the company may continue to sell off subdivisions and retreat to core activities to make it more attractive to future investors.

Organisationally, privatisation of POL has both negative and positive aspects when compared with other industries. Some aspects such as its capital intensive nature may inhibit privatisation, whilst others, such as its presence in international markets and its many saleable assets, are highly attractive. Generally small businesses are easier to sell, whilst shipping may often be a special case in Poland, East Europe and beyond.

The research also raised technical issues and the implications of change for the industry. Poor technology can hold back new ideas, plans, training and skills, for example in terms of logistics, speed or costs. There are conflicts here between state withdrawal, subsidy redirection and market needs, and questions as to when, if and how the issues will be resolved.

All areas of commercial operation are likely to be affected if Poland's 1994 application, results in full or partial membership of the European Community. The underlying principles of harmonisation and liberalisation will translate to POL as subsidy removal, reduced regulation, and open markets, to name only a few effects. More specifically the adoption of Community shipping policy will have implications for the environment, safety, conference membership, flag availability (eg EUROS), cabotage impacts and so forth. These are expansive areas requiring much more detailed work. Meanwhile the development of the TAA provides further grounds for research especially in the context of European Community policy towards conferences, consortia and competition.

POL will also be affected by moves in the market place from East to West. The new openness of markets in the East should be considered in tandem with its recession and instability problems. Notably POL has moved its base port to Hamburg which will have ramifications for Gdynia, including changes in trade direction, new transit routes to the CIS, and possible changes in the ratio of shipping to trucking or rail.

Commercial considerations must include managerial issues, for example whether POL are coping with management change and what real changes have taken place. Requirements for training or fundamental cultural change are also relevant. Industry's changing logistical requirements raise questions as to whether POL can hope to keep up, or whether it will increasingly become a flag of convenience, characterised as a cheap, low standard, unstructured member operator.

Theoretical

Part of this research was aimed at modelling the relationship between East European change and East European shipping. The level of success achieved here must be noted, and the issue raised as to whether a display technique is sufficient, and whether a more quantitative technique could have accommodated the diverse range of issues in a meaningful and substantive way. One of the benefits of the technique chosen is that it is transferable and could be applied elsewhere in East

Europe or beyond. An alternative may have been to create a series of separate sub models, for example one economic, one political and so forth, but it might be questioned as to whether the impacts could then be aggregated to provide an overall assessment, in meaningful terms, of the complex relationships that exist.

The practicalities and value of the contextual model should be considered, as to whether it was useful, comprehensive, transferable and whether it could have been carried out on a more quantifiable basis and still have some meaning. Notably there is scope for extending the managerial and organisational issues into more specific research areas with the development of more detailed models.

The framework approach to analyse the positional changes of POL and its competitors also needs further consideration, and frameworks as a technique need more assessment, as is evidenced by the lack of research into their application, despite conclusive evidence of their soundness theoretically. Many other commercial issues are multifaceted, and may benefit from a framework approach to study, particularly when looking at their contextual interrelationships in a changing social, economic and political environment.

The same is true of positioning generally. The concept appears to be widely accepted but there remains relatively little research. There is a notable dearth of information on measurement techniques as an alternative to Multi Dimensional Scaling which was unsuitable for this research and likely to be equally unsuitable for a variety of other applications. To date, positioning studies have concentrated upon the attitude of the consumer, not the producer, and have centred upon products, rather than services, which leaves considerable scope for further development. It should be noted here that shipping is a peculiar industry as the producer is often highly divorced from the consumer, for example by agents or conferences which may help to explain shipping's apparent naivety of marketing issues. In parallel with this, the size and variety of the industry means there is also scope for further development of the framework model for shipping.

In addition more work is needed on marketing in shipping. The shortage of shipping applications amongst a profusion of marketing literature may reflect shipping's naivety of marketing issues as a significant part of commercial activity.

The relationship between the political environment and shipping might be further researched and a series of models developed, as shipping has a political role to play as a sizeable and strategic industry. These ideas can be expanded to include the development of, for example, legal or social models with shipping, for example to explore the relationship between unemployment/ poverty and shipping industry development in East Europe and elsewhere.

The need for more shipping marketing research can be expanded to include product differentiation, buyer behaviour, pricing practices and effects, image, and the impact of independents operating in competition with conferences.

Moving further away from the specific research analysed here, one of the arguments for conferences which could provide further research areas is that competition often leads to monopolies or oligopolies as the larger operators take over the market, leaving only small niches for the smaller operator. This conflict with conventional competition theory could provide fertile ground for further work.

Finally, a specific paradox stands out from the research. Areas such as marketing, logistics and training, need to be improved in East Europe before any significant profits can hope to be achieved by the old established state firms such as POL, but the same lack of profit means that no finance is available to improve these areas, and lack of initiative may prevent its development in any case.

In highlighting the commercial and theoretical issues, we have shown that in attempting to achieve its objectives the research has covered a wide and diverse range of topics, many of which have presented potential areas for further research.

However, it is hoped that this research has made some progress towards developing and adapting a series of methods which can take this variety of multifaceted information into account when establishing and assessing company position, even against a background of intense instability.

Postscript

In May 1995 the situation regarding the TAA, now TACA (Trans-Atlantic Conference Agreement), remained unresolved. Although the agreement, which now had sixteen member lines, had been banned by the European Commission in October 1994, the European Court of First Instance suspended this ban in March 1995. This meant that there was now a moratorium on prohibitions affecting conference operations on the North Atlantic at least until the court ruled on the overall legality of the TAA. The Commission was free to appeal against this decision but only on a point of law.

Meanwhile, POL withdrew and subsequently sold, four 1,700 TEU con-ros from the North Atlantic line in 1993 in favour of slot charter agreements with Atlantic Container Line and Mediterranean Shipping Company. Two new 3,000 TEU vessels were envisaged for the trade in the near future, finance permitting.

More generally, Poland progressed positively towards economic stability, and confidence was demonstrably returning. Many feel that Poland will eventually take its place among the mainstream European economies, in all probability become a member of the European Union, and be amongst its leaders in the maritime economy.

Appendix 1

Shipping abbreviations used in the text

AN	barge	PB	push barge
AR	crane/derrick barge	PM	passenger/ train/ vehicle vessel
BA	barge carrier		
BB	break bulk	PU	cruise vessel
BC	bulk carrier	PV	passenger/car ferry
BN	bulker	RC	ro ro and cellular space
BO	ore carrier	RF	reefer
BS	bulk carrier ore strengthened	RH	refrigerated fish carrier
		RR	ro ro
CC	converted to FC	SC	semi container ship
CL	container liner	SP	self propelled barge
CN	container ship	TB	tow barge
DC	multi purpose	TC	chemical tanker
DK	livestock carrier	TN	tanker
DN	dry cargo ship	TO	crude oil carrier
DR	part refrigerated	TQ	chemical/oil tanker
GT	trailing suction dredger	TR	products tanker
FC	fully cellular	WA	ro ro
FT	fishing vessel	WB	ro lo
JN	training vessel	WL	ro ro/cellular
MB	ore/bulk/oil carrier	ZO	oil drilling rig

Sources: Containerisation International 1994 and Fairplay World Shipping Directory 1993

Appendix 2

Example of interview structure

Explanation - authors / research purpose
Introduction - basic company information - data, policy, structure
                                            any published material available,
                                            brochures / schedules etc

Details:

Price
Rebates - deferred / %
Discounts - details
Price for transporting one TEU
Market Segmentation - how
Subsidy - amount / type
Currency Convertibility
State Control
Profits
Competition

Promotion
Sponsorship and Hospitality
Advertising - budget, press, types, frequency
Image
Representatives - number, locations
Market Research - budget, people
Public Relations - budget, items
Directories
Literature - types, destinations
Agencies - type, budget

Exhibitioneering
Press Releases

Physical Evidence
Containers - own/lease, own colours
Vehicles
Company Offices
Ports Served
Routes Served

Process
Conference and Consortia - TAA membership
Guaranteed traffic
Collaboration - with

Place
Origins and destinations served
Computer reservation systems

People
Qualifications and training
Employees - number sea/shore
Management structures
Unions/workers councils - membership, role

Product
Speed - kts and days
Safety
Reliability
Frequency - services on North Atlantic
Quantity - TEU capacity
Ship types
Ship ages
Flag
Cross trades
Computer tracking for containers
Trading terms

Check all points covered
Any other points/ further information ?
Thank you

Appendix 3

The European Community shipping policy and East Europe

The important issues here are the shipping policy of the European Community, particularly its consideration of East European competitors and its impact upon European Community liner operations, and to place this situation into context, the development of relationships between East Europe and the European Community in the maritime sector.

With the changes taking place in East Europe, the attitudes of the European Community towards the East have altered considerably. Merritt (1991) commented that the European Community nations' policies towards East Europe might be described as 'chaotic and unpredictable', and that establishing a durable and consistent relationship between the European Community and East Europe is essential although it will not be easy. This section aims to examine the development of a Community shipping policy in the light of these changing attitudes and relationships, and to look at the impacts upon East Europe.

When the Community was founded in the late 1950s, the Soviet Union saw it as a reinforcement of the capitalist camp, detrimental to Soviet interests, and having decided not to accord it juridical recognition continued to treat it for over two more decades with varying degrees of coldness. The Community for its part paid only marginal attention to East Europe throughout the 1960s although as detente began to replace the cold war, European Community members sought to outdo each other both economically and politically in their relations with the East. At first the British set the pace with a fifteen year credit for the Soviet Union in 1964. But the most important of this series of agreements was the Moscow Treaty between the Soviet Union and the Federal Republic of Germany, signed in 1970 which led to more normal relations between the two parties (Pinder 1991). Meanwhile the Treaty of Rome signed on 25th March 1957, under which the EEC was formed, had made no special provision for relations with the states of East Europe, apart from the very specific 'Protocol on German international trade and connected problems'.

Transport and Shipping policy of the European Community

According to Arbuthnott and Edwards (1979) the general aims of the Treaty of Rome included

1   Drawing the people of Europe closer together.
2   Encouraging economic growth.
3   Improving living and working conditions.

and these were to be achieved through the broad policies of harmonisation and liberalisation aimed at creating a level playing field for economic activity in the European Community characterised by a free market framework.

The treaty specified three areas as being essential to unification of the Community; these were:

1   Agriculture.
2   Social Services.
3   Transport.

As far as transport was concerned, the Treaty stated that

the transport market must be organised in accordance with a market economy. Public intervention should occur only where it is otherwise impossible to proceed. The Community must ensure that restrictions to freedom to provide services are removed. At the same time, the aim is to harmonise the overall framework in which modes and companies operate. Therefore we must not lose sight of the objective of optimising the transport process with a view to increasing competitiveness of the EEC, and improving services to the public (Treaty of Rome).

Article 84(1) of the Treaty excluded sea and air transport from the common transport policy, stating that 'the provisions of this title (i.e. Title IV, Transport, of Part Two, 'Foundations of the Community') shall apply to transport by rail, road and inland waterway'. Article 84(2) adds: 'The council may, acting unanimously, decide whether, to what extent and by what procedure, appropriate provisions may be laid down for sea and air transport' (Seefeld 1977).

Hence, for many years, the European Community did not have a shipping policy as the need for one was considered to be excluded by the Treaty. Also the law making process of the European Community was bureaucratically slow and shipping was not as prominent an issue with either governments or the electorate as numerous others such as agriculture or industrial development.

The result was that between 1957 and 1977 there was virtually no legislation relating to shipping. The first moves stemmed from when the UK, Denmark and Ireland, each major seafaring nations, joined the European Community in 1973, and in particular when the Commission backed by the UK took the European Council of Ministers to the European Court of Justice over a narrow maritime issue, relating to the employment mobility of European seafarers, and won. Meanwhile, during the 1980s, the European Community registered fleet declined drastically compared to world fleets, and the need for a stated policy became more than evident (Commission for the European Communities 1989).

Between 1977 and 1985, in particular, several different shipping policy areas were investigated and developed, although little legislation was agreed. These policy areas can be summarised as follows:

1 Social issues relating in particular to free movement of people - including seafarers' employment - between countries, harmonisation of seafarers working conditions, and a range of social security measures relating to the maritime sector, for example social and health benefits.
2 The right to establish shipping companies throughout the European Community following the principles of harmonisation and liberalisation: The Treaty made it illegal for countries to show bias against any non-national attempting to set up a shipping company.
3 Liner shipping conferences were recognised as violating the Treaty of Rome as they were anti-competitive. However, they presented strong political arguments for their retention and hence no progress was made to 1985.
4 Safety and pollution legislation in relation to shipping was agreed before 1985. The European Community entered into negotiations with organisations such as BIMCO, and between 1978 and 1982, produced shipping related Directives, for example concerning pilots and tanker operations.
5 The 'Brussels package' agreed by all member states in 1979 stated that ratification of the UNCTAD code by European Community members was a requirement. The 40/40/20 agreement was to apply to Member States trade with developing countries. Cooper (1977) pointed out however that the code did not cover the activities of outsiders who do not belong to the conferences and hence would not affect operations of CMEA countries.
6 The Commission also recognised that the Community shipbuilding industry was facing economic difficulties and accepted that member states might have to provide subsidies. Although the Community has consistently been anti-subsidy in stated policy, a fixed but reducing

percentage of ship cost was allowed to be subsidised.

7   In 1977 the European and Social Committee for the European Communities (ESCEC) produced an opinion on the 'EEC's transport problems with East European countries'. The report pointed out that the steadily mounting competition from the Eastern bloc in the field of maritime shipping was a cause for grave concern on account of the conditions under which it was flourishing. It added that because they were able to operate freely in the West, Eastern bloc countries were succeeding to an increasing degree in changing the pattern of East-West goods traffic in their own favour. Accordingly, the ESCEC pointed out that not only may Eastern bloc carriers' penetration of the markets threaten employment in transport, but in the long run, there may be grave drawbacks for industry in the Community as a whole. For this reason the Committee called on all the institutions responsible for East-West transport questions to tackle this matter with the utmost vigour in order to ward off developments that would be disastrous for the economy and have grave social consequences (ESCEC 1977). The Committee then outlined the objectives they considered important in the specific field of sea transport as follows:

a   CMEA countries should be made to drop freight rates that are in no balanced relation to the normal terms in Western countries.

b   Community shipowners should be given a balanced share of bilateral traffic between Community and CMEA ports in both directions, at adequate rates and without carriers from other countries being excluded.

c   West European shipowners should be given the chance to acquire a share of traffic between CMEA ports and ports outside the Community.

d   Eastern bloc shipping lines should be allowed to accede to existing agreements between Western shipowners.

The ESCEC along with the Commission and the European Parliament had shown an interest in Eastern bloc activities for a number of reasons as it had created problems for European Community shipping since the end of World War II. From 1977 onwards, the Seefeld Reports investigated the situation and reported back to the European Parliament and Commission - Eastern bloc shipping at this time was described as having a conventional purpose of serving its own markets and as performing dubious defence exercises (Seefeld 1977 and 1979); all vessels were thought to have a defence bias or alternative defence purpose.

The first Seefeld report (23/03/77) called for the Commission to establish a common position toward state trading countries and other countries that wanted to build up their own merchant fleets. It suggested that pressure from outside the Community was forcing Europe to act, and highlighted the fact that state trading countries were 'forcing their way onto the world shipping market and endangering the member states shipping industries'. Until this time the Community had ignored these countries which in default of a clear cut Community decision on maritime policy were able to insist that all their export transactions were effected on a CIF basis and all their import transactions on an FOB basis, which left sea transport completely in their hands and eliminated European shipowners from this trade.

Even more important to the shipping industries of the Community countries was the threat represented by the wide-spread practice adopted by the state trading countries of undercutting conference tariffs in transport operations direct to the member states and especially in cross trades. Many Western shipping companies were convinced that undercutting to an extremely low level in this way corresponded to dumping and pointed out that by Western standards the Eastern bloc fleets were operating uneconomically in that, for example, vessel replacement costs were not being covered, although many of these arguments were disputed by a number of commentators including Bergstrand and Doganis (1987).

These points had already been identified in 1976 by Prescott (1976) who also noted the problematic practices of inter-governmental (either bilateral or multilateral) agreements reserving part or all of the cargo moving in the trade, and the establishment of joint shipping agencies in foreign countries without reciprocity; Western shipping businesses were not afforded the same possibility to run their business in Eastern bloc countries with the same freedom (Russell 1975).

The Seefeld report carried on to point out that state trading countries were 'double-dealing', for example in connection with the code of conduct for liner conferences. On the one hand they voted with the developing countries in UNCTAD when it was a question of 'thwarting the interests of the traditional maritime nations', on the other hand they adopted the position of an outsider by undercutting both industrialized and developing countries' shipping companies.

Seefeld suggested that if the Community did not develop a coherent sea transport policy, the goods to be shipped would always be regarded as more important than the interests of those shipping them. Although the report did not suggest any policy alternatives, it did support the call from the shipping companies associations for effective protection against the non-commercial practices of the state trading countries and hoped that the Community would raise shipping questions in any negotiations between the EEC and the CMEA. Cooper (1977) went on to suggest that the initial aims of a Community shipping policy should include negotiating as a Community with the CMEA for the preservation of a co-ordinated seaborne trade. The second Seefeld report (5/1/79) revealed that the European Community was under pressure of time in working out and implementing a common transport policy from CMEA countries which were 'anxious to push their way on to the world market with its rich pickings in foreign currencies' (Seefeld 1979).

The advance of industrialisation in the CMEA countries was seen as the reason for their emergence on world markets. The state trading countries penetrated world transport markets using all the resources at their disposal:- for example at the cost of the standard of living of their citizens and without regard for profitability. As most Eastern bloc shipping companies were state owned they did not need to operate commercially, only to earn hard convertible, currency without which they would not be able to purchase vital foreign commodities from the West. Being able to charge less than cost because of state subsidy, these vessels were able to undercut western shipping. Although this was seen as unfair trading by the European Community, little action could be taken, as Community countries had no jurisdiction over East Europe, and could not risk taking action which might provoke retaliatory measures from East European countries. The European Community did however set up a review of activities in the international trading and shipping industries of the CMEA with which it also entered into negotiations.

The Seefeld reports suggested that the European Community needed to find ways to persuade the state trading countries to adopt a more western approach without destroying the foundations of free competition or forfeiting the high degree of efficiency that had characterised world sea transport.

Overall, until 1985 there was little shipping policy, and in particular legislation, to show for 28 years of the European Community, although it is interesting to note the significance attached to East Europe throughout this period. The 1970's saw the beginnings of slow talks between the CMEA and the European Community, and by the mid 1980s, the Community's relationship with the CMEA countries seemed to have stabilised at a modest level of activity. In June 1984 the CMEA summit meeting expressed its interest in a relationship with the Community, and in October of the same year a communication from the CMEA to the Community suggested negotiations for an agreement, declaration or other document (Pinder 1991). By the time the first real European Community shipping legislation was developed changing circumstances had caused East Europe to decline in its significance to the Community as a maritime competitor, and other issues became dominant.

Stage 1: shipping policy

To obtain a full picture of the operating circumstances of Community shipping companies, we need to go on to examine the development of policy over the 1980s and 1990s. On 16th December 1986 the European Community, at a meeting of the Ministers of Transport, agreed a maritime package which combined with measures adopted earlier and outlined above formed the basis of a Community Shipping Policy. The package included four regulations outlined by Erdmenger and Stasinopoulos (1988) as follows :

215

1. Council Regulation (EEC) No. 4055 / 86. Freedom to provide services

This introduced the principle of freedom to provide services to intra community trade, distinguishing existing arrangements from future agreements. Applying to nationals of member states, it aimed to prevent any member state from discriminating in favour of its own shipping companies over those from another member state.

It called for unilateral cargo restrictions by member states to be phased out by 1st January 1993. Other discriminatory cargo sharing arrangements between member states should be phased out or adjusted to comply with Community legislation. This regulation was of note to East European countries, as the council could extend it to nationals of a third country who provide shipping services and are established in the Community, for example Polish Ocean Lines' activities into and out of the UK.

2. Council Regulation (EEC) No. 4056 / 86. Competition rules

This aimed to apply Treaty competition rules to shipping, and affected all international shipping services to and from Community ports, except tramp services.

It exempted liner cargo conferences from Treaty provisions on restrictive practices. Restrictive practices of member states may prompt action from the Commission, although Council authorisation was required to deal with conflicts in international law. This had little, if any impact for East Europe as the shipping companies of these countries were usually non conference members. However, it was to be significant in its impact upon the development of the TAA, during 1991, and its subsequent operation, and consequently POL's activities on the North Atlantic.

3. Council Regulation (EEC) No.4057 / 86. Unfair pricing practices

This regulation applied to liner trades, and enabled a compensatory duty to be imposed on non EEC shipowners by the Community if the following conditions were cumulatively present:

i   There are unfair pricing practices - defined as undercutting Community shipping services, where this is made possible because the non Community shipowners enjoy commercial advantages such as subsidy.
ii They cause financial or commercial injury to Community members.
iii The interests of the Community make intervention necessary.

This regulation was particularly relevant to East European shipping companies who were believed to be undercutting Western companies by charging less than

216

cost because of state subsidy (Cooper 1977).

4. Council Regulation (EEC) No. 4058 / 86 Co-ordinated action

This regulation dealt with distortion of competition by governments giving preferential treatment to their own shipowners, and provided for co-ordinated Community action where third countries restrict access of European Community shipping companies to ocean trades.

It was based on the assumption that it is in the interests of Community shipping not to encourage a protectionist approach in the rest of the world. The Community aimed to continue a commercial regime by taking action against non commercial attacks upon it. It is possible that this may have had some East European impact, especially where East European shipping companies had offices in the European Community as well as in East Europe.

Although Stage one was somewhat limited in scope it nevertheless provided the first substantive legislative basis for a European Community shipping policy. By the time Stage two of a shipping policy for the European Community emerged, changes in East Europe such as the accession of Gorbachev, decreasing Soviet intervention in other East European states and declining East European economies meant that shipping in East Europe was no longer of such significance to the European Community, although it was beginning to be replaced by a need to adopt measures to encourage and support the East Europeans' development in the new social, economic and political context.

Stage 2: shipping policy

On 5th June 1989 the Commission produced two separate documents which constituted the basis for stage two of the European Community shipping policy, which were intended should be translated into legislation.

PAPER 1. 'A Future for the Community Shipping Industry : Measures to improve the operating conditions of Community shipping' (Commission of the European Communities 1989)

Several policy initiatives were proposed in this paper:

1   A Community ship register: EUROS
It was hoped that this would be introduced on 1 January 1991, for vessels less than 20 years, and greater than 500 grt. Run in parallel with national flags, EUROS would offer benefits such as:

- easier movement of ships between member states ie technical compatibility

217

- mutual recognition of seafarers qualifications
- the 'opening up' of cabotage to all vessels on the Community register.

Vessels would be required to meet safety and certification levels of every member state; additionally all officers plus half the crew had to be Community nationals. The aims of the scheme included the revitalisation of the European Community national fleets.

Following some slight delay, it was expected that EUROS might come into effect in June 1991 (Lloyds List 16/1/91), but disagreements over the issue of tax relief led to further delays. On 19th November 1991, Lloyds List reported that the European Commission would adopt fresh proposals for a European ship register by the end of the year pending final agreement on the new income tax relief measures for seafarers. By December 18th 1991 European Transport Commissioner Karel Van Miert said his latest amendments to the Commission's proposed EUROS ship register, giving income tax concessions to seamen on EUROS registered ships, had been welcomed by a majority of transport ministers. Despite continuing discussions, no agreement on EUROS had been reached by the beginning of 1994.

2   The Commission suggested a research fund to decrease manning requirements on ships and improve competitiveness.

3   Technical standardisation of equipment for member states, in order to reduce existing conflicts. This related to the overall Community theme of harmonisation.

4   Social measures to improve working conditions, for example decreased hours, common training schemes, and mutual recognition of qualifications. Again this related to harmonisation.

5   Environmental action to ensure the same standards of marine pollution prevention in all ports.

6   As an incentive to join EUROS, all vessels registered under the scheme would receive priority to transport surplus foodstuffs as Community aid.

7   A Community shipowner was defined - an important issue to clarify as it affected cabotage rules.

8   Removal of restrictions on cabotage, which related to liberalisation. Although it attempted to open up the maritime market this was a highly politically contentious area. The Commission put forward the idea that all Community countries should open up their domestic shipping to EUROS vessels. Manning requirements would be the same as for the member state in question, which might

conflict with EUROS requirements. It was proposed that this item be reviewed in January 1993. A 'get-out' clause was provided in that each member state could define specific routes which required subsidy to operate and were needed for public service reasons. Subject to Commission approval, these routes could then be reserved for national carriers.

In December 1990 European Community transport ministers agreed that a first phase of liberalising cabotage was to begin during 1993, amid strong opposition from the Greek and Italian governments and doubts over its legality. The ministers agreed that the first phase should cover mainland cabotage and a second phase island cabotage. Legal doubts surrounded the setting of a date 'during 1993' - after the 1992 deadline, stemming from the Single European Act - and the way that only 'mainland' shipping would be included in the first phase of liberalisation.
In February 1991, lawyers confirmed that the agreement reached in December was in breach of the Treaty of Rome (Lloyds List 20/2/1991). However, the Commissions' lawyers concluded that derogations from a January 1993 cabotage introduction could be made.
In December 1991 Lloyds List reported that European Community transport ministers had agreed a framework for liberalising maritime cabotage and that they hoped to finalise a deal within the following six months. The new scheme envisaged liberalisation by January 1st 1993 of mainland cabotage in the tramp and liner sector, where a journey is seen as one leg of an international voyage. The first stage, from January 1993, would involve liberalising cabotage for tramp shipping and liner traffic between mainland destinations. The second step would involve all other traffic between mainland ports. The last three stages cover island traffic and would begin with tramp and other liner traffic, before all remaining island traffic except passenger traffic, with the last stage being regular passenger ferries (Lloyds List 02/12/91).
However, Lloyds List (24/06/92) reported that the twelve member states were at last in general agreement that for most services crew would be dictated by flag where vessel size was equal to or greater than 650 tonnes. Exceptions included some Mediterranean island routes where higher numbers of nationals were allowed. New cabotage target dates were set in 1992, which would see legislation coming into force throughout the European Community by 2004. The majority vote needed was gained from nine of the member states, those not in favour were UK, Denmark and the Netherlands - who abstained. Criticisms included the number of exceptions and derogations.

PAPER 2. 'Financial and fiscal measures concerning shipping operations with ships registered in the Community' (Commission for the European Communities 1989).

Whereas the first document set out proposals for legislation, this second document was mainly advisory and concerned subsidy, its definition in a European context and the limitations acceptable to the European Community.

The Commission recognised that Community shipping was already heavily subsidised, and that this was distorting the market. However, the Commission agreed to approve certain limited subsidies to help to retain Community ships under Community flags and employ Community seamen. Subsidies were only to be allowed for social security payments, training and differential tax regimes.

On 15th April 1991 the Community drew up 'Guidelines for the examination of state aid to Community shipping companies' (Commission for the European Communities 1991). These guidelines, which replaced the second document of stage 2 of the European Community shipping policy, defined ten types of vessel which could receive different levels of subsidy. The aim of the plan was to restrict subsidies to the lowest levels possible.

The stage 2 package was only a set of proposals which it was hoped, would be implemented by 1993. Whatever other components of the single market were in place by the end of 1992, a common European shipping policy was present only in skeletal form (Lloyds Shipping Economist 1990). Universal agreement on policy remained difficult because of the varying interests of member countries.

A later paper outlining a new European Community maritime policy was drawn up by Dr Bangemann, Vice President of the European Commission. According to Bangemann the advent of the single market within the Community will double the present volumes of cross border transport - the vast majority of it by sea - by the year 2000. At the same time international trade, 90 % carried by shipping, was expected to continue growing (Lloyds List 29/8/91). To prepare the Community for the upturn Bangemann advocated a European Maritime Agency, primarily as a forum for promoting co-operation between all parts of the industry and Community governments. He dismissed previous maritime policy as 'out of date' (Lloyds List 30/8/91).

In an interview with Lloyds List on 3rd September 1991 Dr kroeger managing Director of the Association of German Shipowners, stated that the test for an effective European maritime policy is 'whether it offers conditions which would make the industry want to stay in Europe'. Although commenting that Bangemann's proposals were welcome he felt that they had not gone beyond a 'very initial phase'.

On September 28th 1991 it was reported in the Telegraph that the European Commission had set up a forum to examine Community maritime policies and to find ways of making European Community shipping, shipbuilding and maritime services more competitive. The forum was asked to produce a report and recommendations by summer 1992 (Telegraph 28/9/91).

Meanwhile consortia, which were first raised as a policy issue in 1984, were defined by the Commission as:

coalitions of several independent shipping lines seeking some form of cooperation in order to maintain profitability through rationalisation in the widest service and to spread the expense of investment in container operations.

The Commission concluded that consortia are legally different from conferences so that regulation 4056 does not apply, and that they need to seek block exemption otherwise they are illegal. The industry generally disagrees with this, arguing that consortia allow shipping to compete worldwide and are therefore commercially necessary, and that many are only of a technical nature.

The Commission eventually decided that a block exemption was necessary as overall consortia do provide benefits, such as regular sailings and economies of scale. Hence the enabling regulation 479/92 (25/02/92) which permitted the Commission to adopt in due course, detailed rules for the automatic exemption of Consortia from Article 85(1) of the Treaty of Rome.

On 15th November 1993 Lloyds List reported details of the draft, which stated that any consortia with a 50% or higher market share had to notify the Commission of their agreements for individual approval. Conference member lines were allowed 30% of a market and non conference members 35%, or 50% with special agreement. The rules allow fixed timetables, exchange slots, pooled capacity, joint offices and terminals and pooled receipts/tonnage. Liner companies wishing to join a consortium would be limited to a maximum 18 month period after which they can leave the consortium although there would be a six month notice period of leaving. The market share and joining regulations were unpopular with the shipping industry, who stressed the already highly competitive market environments and the need for greater commitment to ensure consortia success.

The developing consortium regulations are clearly relevant to the design and future of the TAA, and thus to this research in particular in that Regulation 4056 does not apply, but by the beginning of 1994 these regulations had still not passed into law.

New members

The majority of East European countries have aspirations of joining the Community. Their hopes were heightened in May 1991 with a suggestion from within the European Community executive that a form of flexible affiliation could solve many of the problems blocking membership. The idea suggested that criteria for membership must be loosened if they were to be appropriate for developing economies (Lloyds List 05/91). On 16th December 1991 Poland entered into an associate agreement with the Community (EIU 1992). However, Community sources described its timetable as over optimistic and indicated several problems that would need to be dealt with before associate status was possible. Overall, Polish membership of the European Community would still appear some way off.

Perhaps its only impact for this research is in terms of the Polish economy adapting slowly to European Community needs, as noted in the Contextual model.

Since the recent economic and political changes have taken place in East Europe, maritime developments in that area have received little attention from the European Community. Whereas in the early stages of formulation of a maritime policy, the Community saw East European countries as a competitive threat (ESCEC 1977) this is no longer the situation. The main concern of the Community now is how to help former members of the CMEA survive the difficult transitional period. It is in the interests of the Community that pluralist democracy and the market economy should prevail throughout Europe (Pinder 1991), and the Community must consider what it can do to help East Europe towards this situation, and how in this process to aid East European shipping as part of economic redevelopment. In the meantime, other issues have taken a greater role, for example the SEM, harmonisation and liberalisation, and European Monetary Union, and these temporarily have pushed European maritime issues back.

Conclusions

Since the early days of the development of a European Community shipping policy, when concern about East European shipping activities was evidenced by the reports of Prescott (1976-77), Cooper (1977), ESCEC (1977) and Seefeld (1977 & 1979), it is noticeable that East Europe has declined in its significance as other issues have become dominant, and the recent dramatic changes in East Europe, for example the accession of Gorbachev and the collapse of the Council for Mutual Economic Assistance (CMEA), have called for a change in the attitude of the Community towards East Europe generally, as well as specifically in relation to shipping issues.

Looking at the European maritime scene, an article in Lloyds Shipping Economist, February 1990, highlighted that 'it is likely that the main issues will remain unresolved for a long time yet' (Lloyds Shipping Economist 1990). This indicates that the situation regarding the relationship between European Community and East Europe and their shipping policies is likely to remain unsettled for the time being. Although European Community progress towards a common policy has been slow, Stage three of a Community shipping policy seems likely to emerge eventually; the effect that this has on East Europe will depend upon whether East European countries decide to join and are accepted by the European Community. According to Merritt (1991) the East Europeans eventual membership of the Community is inevitable. During his Commission Presidency, Jacques Delors repeatedly emphasised the point to political leaders across Europe that if the rest of the Community wants Germany to remain firmly anchored inside the Community, Eastern Europe cannot be left outside it. The turning towards the West of the former CMEA countries marks a fundamental shift in relationships within Europe, which will undoubtedly have impacts upon the shipping industry.

Although East Europe has declined in its significance to the European Community recently in terms of shipping policy, a number of major issues in this area will remain of interest to the Community. These may include the consideration of Western access to East European markets. Also there are still likely to be some concerns about East Europe's need to earn hard currency and the methods by which this is achieved; for example by undercutting western shipping operations. Linked to this are concerns over hidden maritime subsidies in East Europe leading to anti-competitive practices. The Community may also be interested in the advance in technology and efficiency of East European shipping industries which may be achieved through the introduction of privatisation or joint ventures. These broad maritime issues are likely to remain of some interest and concern to the European Community until a better understanding and a closer relationship develops with East Europe.

Finally, the development of a European Community shipping policy in general has a clear significance for liner operators, including those of East Europe. As a consequence it is of direct relevance to this research.

References

Ackoff R.L. 1962 *Scientific Method: optimising applied research decisions* Wiley, New York

ACL/MSC joint service press release 1992

Adams G.R. and Schvaneveldt J.D. 1991*Understanding Research Methods* Longman

Ambler J, Shaw D.J.B. and Symons L. 1985 *Soviet Union and East European Transport problems* Croom Helm

Arbuthnott H. and Edwards G.1979 *Common Mans Guide to the Common Market* Macmillan Press

Babbie E. 1992 *The Practice of Social Research* Wadsworth Publishing Company

Bagozzi R.P. 1994 *Principles of Marketing Research* Blackwell Publishers

Baker M.J. 1991 *Marketing: An introductory text* 5th Edition, Macmillan Press

Baker T.L. 1994 *Doing Social Research* 2nd Edition McGraw Hill

Barro R.J. *Macroeconomics* 3rd edition, John Wiley and Sons Ltd 1990

Beck P.W. 1983 *Forecasts: opiates for decision makers* a lecture to the third international symposium on forecasting, 5th June, Philadelphia

Beenstock M. and Vergottis A. 1993 *Econometric Modelling of World Shipping* International studies in Economic Modelling Series 16

Berstrand S. and Doganis R. 1987 *The impact of Soviet shipping* Allen and Unwin

Bessom R.M. and Jackson D.W. *Service retailing a Strategic Marketing Approach* Journal of Retailing, 51, Summer, pp 75 - 84

Blazyca G. and Rapacki R. 1991 *Poland into the 1990s* Pinter Publishers, London

Booms B.H. and Bitner M.J. 1981 Marketing Strategies and Organisational Structures for service firms, in Donnelly J. and George W.R. (eds) *Marketing for services* American Marketing Association, Chicago, pp 47 - 51

Borden N.H. 1965 The Concept of the marketing mix, in Schwartz G. *Science in Marketing*, J. Wiley and Sons, New York, pp 386-397

BOTB (British Overseas Trade Board) 1990 *Country Profile : Poland.*

Bradley F. 1991 *International Marketing Strategy* Prentice Hall International Ltd

British Overseas Trade Board 1989 *Joint ventures in Eastern Europe* August

Brundage J. and Marshall C. 1980 Training as a marketing management tool *Training and Development Journal*, Nov, pp 71 -76

Bryman A. and Burgess R.G. 1994 *Analysing Quantitative Data* Routledge *Business Central Europe* 1993 June p13

BSC 1993 Complaint to the EC Commission 18th December 1992 and *Response to the Requests for the Interim Measures of the British Shippers Council and the Cousiel National des Usagers des Transports - Annex 3: Overcapacity and Capacity Regulation in the Transatlantic Trades* 11/01/93

Cannon T. 1980 *Basic Marketing :Principles and Practice* Holt, Rinehart and Winston, New York

Carden P. 1984 Paper 1 *Liner and through transport operations* Conference on Maritime Joint Ventures, 17th and 18th May, Royal Lancaster Hotel, London.

Centre for Privatisation 1990 *Act on the privatisation of state owned enterprises*, The office of the minister of ownership changes Act, Warsaw, Poland.

Chadwick G.1971 *A Systems view of planning* Pergamon Press, Oxford

Chadwick B., Bahr H.M., and Albrecht S.L. 1988 *Social Science Research Methods* Prentice Hall

Chmara E. and Langley P.E. 1973 *Evaluation matrices for structure plan* (Department of the Environment; London)

Clark B.D., Gilad A., Bisset R. and Tomlinson P. (eds) 1984 *Perspectives on Environmental Impact Assessment* proceedings on annual training courses on EIA sponsored by the World Health Organisation, D.Reidel Publishing Company

Clarke D. and Rivett B.H.P. 1978 *A structural mapping approach to complex decision making* Jl Opl Res Soc 29(2) 113-128

Clarke M. and Herington A. 1988 *The role of Environmental impact assessment in the planning process* Mansell Publishing Ltd.

Coastal Times 16/10/94 Reflagging the Polish Merchant Fleet No. 10

Commission for the European Communities 1989 Paper 1 *A future for the EC shipping industry: Measures to improve the operating conditions of Community shipping* COM (89) 266 Final 3 August

Commission for the European Communities 1989 Paper 2 *Financial and fiscal measures concerning shipping operations with ships registered in the Community* SEC (89) 921 Final 3 August

Commission for the European Community 1991 *Opening up the internal market*

Commission for the European Communities 1991 *Guidelines for the examination of state aid to community shipping companies*

Containerisation International Yearbook 1991

Containerisation International 1992 CGM's North American retreat June

Containerisation International 1993 *POL revamps strategy* January

Cooper A. 1977 "Shipping Policies of the EEC" Maritime Policy and Management

COS 1990 *Czechoslovak Ocean Shipping Annual Report*

COS 05/02/92 Interview with Podobsky J. (Assistant Commercial Director) and Halouska J. (Planning Department) of Czechoslovak Ocean Shipping, London

Cowell D. 1991 *The Marketing of Services* 2nd Edition Butterworth - Heinemann Ltd.

Cox T. and Cox M.A.A. 1994 *Multi Dimensional Scaling* Chapman and Hall, London

Coxon A.P.M. 1982 *The users guide to Multi Dimensional Scaling* Heinemann Ltd.

CSX Corporation 1992 *Annual Report Form 10-K*

Darymple D.J. and Parsons L.J. 1990 *Marketing Management* 5th Edition John Wiley and Sons

Davidson D.S. 1978 How to succeed in a service industry....Turn the organisational chart upside down *Management Review*, April, pp 13 - 16

Davison M.L. 1983 *Multi Dimensional Scaling* Wiley, New York

Day G. and Wensley R. 1988 Assessing advantage: A framework for diagnosing competitive superiority *Journal of Marketing* 52 (April) 1 - 20

Department of Environment Research Report II 1976 Environmental Impact Analysis HMSO

Department of Transport 1977 Oct *Report of the advisory committee on trunk road assessment* Chairman: Sir George Leitch HMSO

Department of Transport 1979 Oct *Trunk Road Proposals - A Comprehensive Framework for Appraisal* The Standing Advisory Committee on Trunk Road Assessment, Chairman: Sir George Leitch HMSO

Dinwoodie J. 1988 *Shipping operational research : in hard cases or soft ?* Plymouth Polytechnic Department of Shipping and Transport

Domoustchiev 26/11/91 Interview with Cpt Domoustchiev - Chairman and Managing Director, Balkan and Black Sea Shipping Company, London

Donnelly J. and George W.R. 1981 *Marketing of Services* AMA, Chicago

Drabnek Z. 1989 CMEA: The primitive social integration and its prospects, in *Economic aspects of regional trading arrangements* Eds Greenaway D, Hyclak T, Thornton R, Harvester Wheatsheaf

Drewry 1993 *Containerisation in the 1990s* Drewry Consultants Shipping Report

Drucker P.F. 1977 *Management tasks, responsibilities, practices* Harpers College Press, New York

East European Economic Handbook, 1985, Euromonitor Publications Ltd, 1st Edition

East European Markets 01/05/92

Economic Review 22/05/92 Polish News Bulletin of the British and American Embassies, Warsaw, Government Privatisation programme - An outline Nowa Europa

Economist 20/04/91 Comecon. Life after death

226

Economist 31/08/91 The Russian Revolution
Economist 1/2/92 Democracy in Eastern Europe
Economist 07/03/92 The people's car
Economist 16-22/05/92 Crossing the East-West chasm
Economist 18/09/93 Poland: Off to the Polls
Economist 25/09/93 Poland: Not as bad as it looked - maybe
EIU (Economist Intelligence Unit) *Poland Country Report* No 1 April 1992
Erdmenger J. and Stasinopolous 1988 The shipping policy of the EC, Common market law review. *Journal of Transport and Economic Policy* September
Ernst and Young 1990 Commission of the European Communities. Analysis of the Maritime Transport Sector of East European Countries. Final Report. Vol 5: *The Maritime Transport Sector in Bulgaria.* November
Ernst and Young 1990 Commission of the European Communities. Analysis of the Maritime Transport Sector of East European Countries. Final Report. Vol 4: *The Maritime Transport Sector in Poland.* November
ESCEC 1977 *EEC's transport problems with East European Countries* Opinion
ESCEC 1990 *Preliminary draft information report of the Section for External Relations,* Trade and Development Policy on Poland 14th May
Europa 1994 *East Europe and the Commonwealth of Independent States 1994* 2nd Edition Europa Publications Ltd
Evans J.J. and Marlow P. 1991 *Quantitative methods in maritime economics* Fairplay Publications
Fairplay 15/03/90 Finns to establish joint venture in USSR
Fairplay 10/05/90 Gdansk yard partially privatised
Fairplay 31/05/90 SeaLand in Soviet joint ventures
Fairplay 12/07/90 Romanian line seeks partners
Fairplay 12/7/90 West Germany cuts yard subsidies
Fairplay 15/8/91 Poland's pace of change is still too slow
Fairplay 15/08/91 Poland p22-31
Fairplay 03/09/92 World in focus: Poland p28
Fairplay 22/10/92 German shipping subsidy ?
Fairplay World Shipping Directory 1987
Fairplay World Shipping Directory 1989
Fairplay World Shipping Directory 1991
Fairplay World Shipping Directory 1993
Farthing B. 1985 Paper 6 Seoul Symposium on Maritime Joint Ventures, Korea
FEARO 1978 *Guide for Environmental Screening* Federal Activities Branch, Environmental Protection Service, Canada
Finance East Europe 1992 May
Financial Times 20/11/90 Survey Poland, Walesa calls for haste
Financial Times 10/05/91 Comecon gives up the ghost of Stalin
Financial Times 29/06/91 Comecon put out of misery after 42 years
Financial Times 24/07/91 Comecon divorce bogged down in squabble over

property

Financial Times 28/10/91 Europe`s trade deal heartens EC applicants

Financial Times 28/10/91 Polish election produces no clear winner

Financial Times 29/10/91 Mazowiecki narrowly ahead in Polish vote

Financial Times 29/10/91 Patience runs out among disillusioned electorate

Financial Times 01/03/92 GM signs joint venture deal with Poland

Financial Times 3/2/92 Polish economy declines despite private growth

Financial Times 6/4/92 Voters get to grips with ballot reform

Financial Times 24/02/93 Poland poised to be a post communist success story

Foreign Trade Research Institute 1992 *Poland Your Business Partner* Ministry of Foreign Economic Relations, Warsaw

Fortlage C.A. 1990 *Environmental Assessment: A Practical Guide*

Friedman M.L. 1991 *The AMA Handbook of marketing for the service industries* Congram C.A. and Friedman M.L. (Editors), AMACOM

Friend J. and Hickling A. 1987 *Planning under pressure* Pergamon Press

Green P.E., and Carroll J.D. 1970 *Multidimensional Scaling and related techniques in marketing analysis* Allyn and Bacon Inc. Boston

Green P.E. and McMennamin 1973 - quoted in Wind Y. 1982 *Product policy, concepts, methods and strategy* Addison Wesley Publishing Company

Green P.E. and Tull D.S. 1978 *Research in Marketing Decisions* Prentice Hall

Green P.E., Tull D.S. and Albraum G. 1988 *Research for Marketing Decisions* 5th Edition Prentice Hall

Gross A.C., Banting P.M., Meredith L.N., Ford I.D. 1993 *Business Marketing* Houghton Mifflin Co.

Guardian 20/12/90 *CMEA replacement put forward*

Guardian 6/2/92 *Recession forces Poland to slow down economic reform*

GUS Statistical Yearbook, Warsawa Oct 1990 *Informacja statyczna o sytuacj spoleczno-gospodarczej Kraju*, Parts 1 and 2

Guzhenko T 1977 Soviet Merchant Marine and World Shipping *Marine Policy* April, 1, 2,

Hill M. 1960 *A method for evaluating alternative plans. The Goals Achievement Matrix applied to transportation plans* PhD Dissertation, University of Pennsylvania

Hill M. 1968 A Goals Achievement Matrix in Evaluating urban plans *Journal of the American Institute of Planners* 34, 2, 19-29, 1968

Hill M. 1973 *Planning for Multiple Objectives* R.S.R.I. Monograph Series r, Philadelphia

Hooley G.J. and Saunders J. 1993 *Competitive Positioning - The key to market success* Butterworth Heinemann

ICC 1989 *A guide to maritime joint ventures* Centre for Maritime Cooperation. Publication 474

Independent 29/06/91 Few tears shed as Comecon is laid to rest

Independent 06/03/92 Polish MPs block economic relaxation

Independent 23/04/92 Sacrifices begin to pay off for Polish economy

Independent 18/05/92 Czechoslovaks take road to capitalism

Independent 29/02/92 Survey finds that Poles are becoming more xenophobic

Independent 31/05/93 Poles move to outlaw coalition nightmare

International Management 1991 November Eastward Woe

International Management 1993 Poland's Economic Miracle October

Jain AK, Pinson C, and Ratchford BT 1982 *Marketing Research: Applications and Problems* John Wiley and Sons

Jain R.K., Urban L.V. and Stacey G.S. 1981 *Environmental Impact Analysis: a new dimension in decision making* 2nd Edition

Jessiman W. , Brand D., Tumminia A. and Brussee C.R. 1967 *A rational decision making technique for transport planning* Highway Research Record no.180, 71 80

Joyce F.E. and Williams H.E. 1976 *Conceptual Frameworks for Environmental Evaluation and Research* University of Aston

Keegan W. 1989 *Global Marketing Management* 4th Edition, Prentice Hall International Editions

Kolankiewicz G. and Lewis P.G. 1988 *Poland: Politics, Economics and Society* Pinter Publishers

Koralewicz J., Bialecki I. and Watson M. 1987 *Crisis and Transition* St Martins Press, New York

Kotler P. 1971 - quoted in Wind Y. 1982 *Product policy, concepts, methods and strategy*Addison Wesley Publishing Company

Kotler P. 1993 *Principles of Marketing* 6th Edition Prentice Hall, Englewood Cliffs, p624

Kotler P. 1993 *Marketing: An Introduction* 3rd Edition Prentice Hall

Kotler P and Armstrong G 1992 *Marketing - An Introduction* 3rd Edition Prentice Hall Inc., Englewood Cliffs, New Jersey

Kreditor A.1967 *The provisional plan industrial development and the development plan*, chapter 8, An foras forbartha; Dublin

Kruscal JB 1964a Multidimensional scaling by optimising goodness of fit to a nonmetric hypothesis *Psychometrika* 29, 1-27

Kruscal JB 1964b Nonmetric multidimensional scaling: a numerical method *Psychometrika* 29, 115-129

Kucharzyk K 1977 *A comparative critique of economic evaluation methods* Msc II Seminar Paper University of Aston

Lancaster G. and Massingham L. 1993 *Essentials of Marketing* 2nd Edition, McGraw Hill

Layard R and Glaister S 1994 *Cost Benefit Analysis* Cambridge University Press, 2nd Edition

Leopold LB, Clarke FE, Hanshaw BB and Balsley JR 1971*A procedure for evaluating environmental impact* US Geological Survey Circular 645, Washington DC

Lichfield N. 1960 Cost benefit analysis in city planning *Jl of the American inst of planners* vol 26 pp 273-279

Lichfield N. 1964 Cost benefit analysis in plan evaluation *Town Planning Review* vol. 35 pp 160-169

Lichfield N. 1966 *Cost benefit analysis in town planning: A case study of Cambridge* Cambridgeshire and Isle of Ely County Council

Lichfield N. 1969 *Cost benefit analysis in urban expansion: A case study, Peterborough* Regional Studies vol.3 pp 123-155

Lichfield N. 1970 Evaluation Methodology of urban and regional plans; a review *Regional Studies* 4 151-165

Lichfield N., Kettle P. and Whitbread M. 1975 *Evaluation in the planning process* Pergamon Press, Oxford

Lloyds List 16/01/91 Euros bids for June start

Lloyds List 20/02/91 EC cabotage plans face a rethink

Lloyds List 13/05/91 East Europeans may be 'affiliated' to EC

Lloyds List 29/08/91 Bangemann sees a new maritime era

Lloyds List 30/08/91 Saving EC industry through growth and co-ordination

Lloyds List 02/09/91 Soviet shipping readies itself for old Union's break up

Lloyds List 03/09/91 Effective European Maritime Policy vital

Lloyds List 10/09/91 Poland Special Report

Lloyds List 12/09/91 Economou Romanian Agency

Lloyds List 11/10/91 Soviet Union Special Report:Shipping Companies use Joint Ventures to halt fleet decline, and, Divergence in face of increased competition

Lloyds List 29/10/91 Soviet Union invites Greek joint ventures

Lloyds List 31/10/91 Romania to upgrade fleet in joint ventures

Lloyds List 01/11/91 Romania banking on joint ventures

Lloyds List 02/11/91 NEC in pact on transport with USSR

Lloyds List 19/11/91 European register set for adoption

Lloyds List 02/12/91 Progress on EC cabotage

Lloyds List 03/01/92 Soviet venture likely to bring owner benefits

Lloyds List 06/03/92 Bulk operators in joint venture

Lloyds List 23 and 24/06/92 EC agrees plan on cabotage

Lloyds List 10/07/92 Wilhelmsen sets up Joint Venture with Poles

Lloyds List 15/01/93 Polish ro-ro capsizes - 50 lost

Lloyds List 15/11/93 Owners blast EC liner rules

Lloyds List 13/01/94 Romania to step up Privatisation

Lloyds Maritime Directory 1992

Lloyds Shipping Economist 1990 EC shipping policy flagging February

Lloyds Shipping Economist 1990 East West co-operation, Aug

Lloyds Shipping Economist 1990 Poland - harsh realities of economic transformation, October, Vol 12, No 10

Lloyds Shipping Economist 1990 Newbuildings Similar problems, contrasting solutions, October

Lloyds Ship Manager 1991 Special Supplement: Reformation in Soviet Shipping August

Lloyds Ship Manager 1991 Greece quick to grasp chances in East Europe December

Lloyds Ship Manager 1992 Special Supplement: Commonwealth of Independent States, August

Lovelock C.H. 1992 *Managing Services* 2nd Edition Prentice-Hall Inc., Englewood Cliffs

Lovelock C.H. 1991 *Services Marketing* Prentice Hall International Inc.

Luck D.J., Wales H.G., Taylor D.A., and Rubin R.S. 1982 *Marketing Research* 6th Edition Prentice Hall

Machota J. 1989 *Czechoslovakia on the High Seas* Orbis Press Agency, Prague

Maslov G.A. 1984 *Development of Soviet Liner Shipping* Institute of Liner Shipping Economics Conference Report International Symposium on Liner Shipping II

Massam B.H. 1980 *Spatial search - Application to planning problems in the planning sector* Pergamon Press Oxford

Massam B.H. and Askew I.D. 1982 *Methods for comparing policies using multiple criteria: an urban example* vol.10 no.2 pp195-204

McAllister D.M. 1980 *Evaluation in environmental planning* MIT Press, Cambridge Massachusetts

McDonald M. 1980 *Handbook of Marketing Planning* MCB Publications 28.

McDonald M. and Leppard J.W. 1992 *Marketing by Matrix* Butterworth Heinemann

McNeill P. 1990 *Research Methods* 2nd Edition, Routledge

Merritt G. 1991 *Eastern Europe and the USSR* Kogan Page Ltd

Mieczkowski B. 1978 *Transportation in East Europe* East European Quarterly, Columbia University Press

Miles M.B. and Huberman A.M. 1994 *Qualitative Data Analysis* 2nd Edition Sage Publications

Millard F. 1994 *The new anatomy of Poland* Gower

Moreby D. 13/12/90 lecture at University of Plymouth

Moreby D. 1991 January, Unpublished paper, University of Plymouth

Morgan N.A. 1991 *Professional Services Marketing* Butterworth Heineman

Nelson D.N. 1992 *Romania after tyranny* Westview Press Inc.

Nie N.H., Bent D.H. and Hull C.H.1975 *Statistical package for the social sciences* 2nd edition McGraw Hill New York

Nijkamp P. 1979 *Multidimensional spatial data and decision analysis* John Wiley, Chichester

North West Water Authority 1978 *Aqueducts and treatment plants* Engineering Report, September

OECD 1986 *Competition Policy and Joint Ventures*

OECD, EBRD, World Bank and IMF 1990 *The Economy of the USSR*

Open University 1989 M205 *Fundamentals of Computing*, Block V Information Systems, Unit 4 Data Modelling

Parker B.C. and Howard R.V. 1977 *The first environmental monitoring and assessment in Antarctica: The Dry Valley drilling project* Biological Conservation 12(2), pp 163-177

Patton M.Q. 1987 *How to use qualitative methods in evaluation* Sage Publications

Payne M.S. 1982 *Individual in-depth interviews can provide more detail than groups* Marketing Today (Elrick and Lavidge, 1, 1982)

Pearce D.W. and Nash C.A. 1981 *A Social Appraisal of Projects: A text in   cost benefit analysis* The Macmillan Press

Persson G. and Backman L. Logistics in Eastern Europe *Logistics and Transportation Review,* Vol 29 No 4 p319 - 334

Pinder J. 1991 *The European Community and Eastern Europe*

Pole Position 1991 October - Newsletter on Privatisation in Poland No 5.

Prescott J. L. 1976 *Community Shipping Industry* EC European Parliament Working Documents

Ramberg J. 1984 *The legal background of Maritime Joint Ventures* Paper 9, Conference on Maritime Joint Ventures

Ratesh N. 1993 *Romania: The entangled revolution* Praeger Publishers

Ries A. and Trout J. 1981 *Positioning* New York McGraw Hill Book   Company

Rivett B.H.P. 1977a Policy selection by structural mapping *Proc. R. Soc.* 354, 407-423

Rivett B.H.P. 1977b Multidimensional scaling for multi-objective policies *Omega* 5(4) 367-379

Rivett B.H.P. 1978 Structural mapping applied to single value policies *Omega* 6(5) 407-417

Rivett B.H.P. 1980 Indifference mapping for multiple criteria decisions *Omega* 8(1) 81-93

Roe M.S. 1980 *Evaluation Matrix Techniques* University of Aston,   Birmingham

Roe M.S. 1991 *The role of Eastern bloc road hauliers in the European Community* Report for the Commission of the European Communities

Russell W. R. 1975 Press release on a speech by retiring chairman of the   Council of European and Japanese National Shipowners (CENSA) in London   on 11/12/75

Rzeczpospolita 01/06/87 Polish Merchant Fleet: Will it survive or Sink ? No 37

Schlager K. 1968 *The rank based expected value method of plan evaluation* p153 Highway Research Record 238, Highway Research Board, Washington DC, 123-152

Seatrade Business Review 1989 Survey September/ October p29-33

Seatrade 1993 April, Containers: Working hard to stand still p47-51

Seefeld H. 1977 *Interim report on sea transport problems in the Community* EC European Parliament Document 5/77

Seefeld H. 1979 *Report on the present state and progress of the common transport policy* EC European Parliament Working Document 512/78

Shepard R.N. 1962a The analysis of proximities: Multidimensional scaling with an unknown distance function I *Psychometrika* 27, 125-140

Shepard R.N. 1962b The analysis of proximities: Multidimensional scaling with an unknown distance function II *Psychometrika* 27, 129-146

Shepard R.N., Romney A.K. and Nerlove S.B. 1972 *Multidimensional Scaling: Theory and Applications in the behavioural sciences* Volume 1 Seminar Press

Shostack G.L. 1977 Breaking Free from product Marketing *Journal of Marketing* 41 (April) 73-80.

Shostack G.L. 1987 Service Positioning through structural change *Journal of Marketing* 51 (January) 34-43

Staar R.F. 1993 *Transition to Democracy in Poland* St Martins Press, New York

Stanton W.J. 1981 *Fundamentals of Marketing* McGraw-Hill, New York, p 446 8th Edition Stanton W.J. and Futrell C.1987

Stewart T. 1981 A descriptive approach to multiple criteria decision making *Jl Op Res Soc* 32 45-53

Stopford M. 1988 *Maritime Economics* Unwin Hyman

TAA minutes of Gulfway meeting 08/10/92

Telegraph 28/09/91 EC maritime forum proposed

The Treaty of Rome 1957

Trout J. and Ries A. 1972 The positioning era cometh *Advertising Age*, April 24, Positioning cuts through chaos in market place, *Advertising Age*, May 1, 1972; and How to position your product, *Advertising Age*, May 8,

Trout J. and Ries A. 1986 *Marketing Warfare* McGraw Hill, New York

Tull D.S. and Hawkins D.I. 1980 *Marketing Research* Macmillan Publishing

Voodg H. 1983 *Multicriteria Evaluation for Urban and Regional planning* Pion Ltd

Walesa L. and TKK Joint statement 1986, Uncensored Poland, October

Walker R. 1985 *Applied qualitative research* Gower

Wall Street Journal Europe 1991 *Privatisation and Market Reform in East Europe and the Soviet Republics*

Walshe G. and Daffern P. 1990 *Managing Cost Benefit Analysis* Macmillan Education Ltd

Westman W.E. 1985 *Ecology, Impact Assessment and Environmental Planning*

White S., Batt J., Lewis P.G., 1993 *Developments in East European Politics*

Whiteman J. 1981 *The Services Sector - A Poor Relation ?* Discussion paper no.8, NEDO, London

Wind Y. and Robinson P.J. 1972 *Attitude research in transition* Russell I.Haley (Editor) Proceedings of meeting sponsored by American Marketing Association

Wind Y. 1982 *Product policy, concepts, methods and strategy* Addison Wesley Publishing Company

World Bank 1992 *Poland* September

Zonis M. and Semler D. 1992 *The East European Opportunity* John Wiley and Sons Inc